# TAUNGYA:
## FOREST PLANTATIONS WITH AGRICULTURE IN SOUTHEAST ASIA

**Sustainable Rural Development Series**

---

# TAUNGYA:

# FOREST PLANTATIONS WITH AGRICULTURE IN SOUTHEAST ASIA

*Interplanting crops can reduce land use conflicts and increase sustainability of forestry in less developed countries*

*Edited by*

Carl F. Jordan
*Institute of Ecology*
*University of Georgia*
*Athens, Georgia, USA*

Jiragorn Gajaseni
*Faculty of Science*
*Chulalongkorn University*
*Bangkok, Thailand*

Hiroyuki Watanabe
*Faculty of Agriculture*
*Kyoto University*
*Kyoto, Japan*

C·A·B International

C·A·B International
Wallingford
Oxon OX10 8DE
UK

Tel: Wallingford (0491) 32111
Telex: 847964 (COMAGG G)
Telecom Gold/Dialcom: 84: CAU001
Fax: (0491) 33508



A catalogue record for this book is available from the British Library

ISBN 0 85198 801 6

Typeset by Leaper & Gard Ltd, Bristol
Printed and bound in the UK by Redwood Press Ltd, Melksham

# Contents

**Part II: Regional Perspectives on Taungya**

# Contributors


Binod Bhatta *Ministry of Forestry, Kathmandu, Nepal.*

L.A. Bruijnzeel *Faculty of Earth Sciences, Free University, De Boelelaan 1085, 1081 HV Amsterdam, The Netherlands.*

Jiragorn Gajaseni *Department of Biology, Faculty of Science, Chulalongkorn University, Bangkok 10500, Thailand.*

Yingqiang Huo *Forest Research Institute of Guangdong, Guangzhou 510520, China.*

Carl F. Jordan *Institute of Ecology, University of Georgia, Athens, Georgia 30602, USA.*

Jay B.S. Karki *Social Science Division, International Rice Research Institute, PO Box 933, Los Baños, 1099 Manila, Philippines.*

O.N. Pandey *Faculty of Forestry, Bisra Agricultural University, Ranchi, India.*

Montee Phothitai *Planning and Statistics Sub-division, Forestry Industry Organization, Bangkok 10100, Thailand.*

Ob Preeyagrysorn *Payap University, Chiangmai, Thailand.*

Hasanu Simon *Forestry Faculty, Gadjah Mada University, Yogyakarta, Indonesia.*

K.P. Singh *Department of Botany, Banaras Hindu University, Varanasi 221005, India.*

Shinya Takeda *Laboratory of Forest Resources, Division of Tropical Agriculture, Faculty of Agriculture, Kyoto University, Kyoto 606, Japan.*

Hiroyuki Watanabe *Laboratory of Forest Resources, Division of Tropical Agriculture, Faculty of Agriculture, Kyoto University, Kyoto 606, Japan.*

K.F. Wiersum *Department of Forestry, Wageningen Agricultural University, POB 342, 6700 AH Wageningen, The Netherlands.*

# Introduction

'Sustainable development' is a phrase popularized by *Our Common Future*, a 1987 report from the United Nations Commission on the Environment and Development. The report defined sustainable development as economic activity that 'meets the needs of the present without compromising the ability of future generations to meet their own needs'. In as much as forests and forest products have played an important role in the economic development of many tropical countries, sustainable forestry is an important component of sustainable development.

In the above context, sustainable forestry could be defined as 'forest management that meets the needs of the present without compromising the ability of future generations to meet their own needs'. The definition seems straightforward, but the question of what needs have top priority is the subject of much debate. Opinions range from timber for international exchange through extractive products for local community use to maintenance of gene pools and biological diversity (Gale and Cordray, 1991).

Regardless of the forest products and values considered to be most important, it is clear that with present rates of deforestation, they are not sustainable. Thailand is a case in point. Between 1971 and 1985, annual production of teak fell approximately 87% (Table I.1). The causes of the decline were decimation of the forests by overharvesting and loss of forest land to agriculture (Gajaseni and Jordan, 1990). There are little data on other forest products and values, but they would be expected to decline proportionally to timber harvest. Clearly, where forests are not sustainable, forestry is not sustainable.

Plantation forests have been one answer to the problem of depletion of wood sources. As natural forests have disappeared, governments and corporations have turned to forest plantations to meet the demand for

**Table I.1.** Teak harvest in Thailand from 1971 to 1985.

| Year | Volume ($m^3 \times 10^3$) | | |
|---|---|---|---|
| | By licence | Confiscated | Total |
| 1971 | 132.7 | 166.1 | 298.8 |
| 1972 | 86.0 | 91.6 | 177.9 |
| 1973 | 165.9 | 28.2 | 194.1 |
| 1974 | 215.6 | 38.7 | 254.3 |
| 1975 | 144.0 | 72.2 | 216.2 |
| 1976 | 170.0 | 93.7 | 263.7 |
| 1977 | 112.1 | 25.8 | 138.0 |
| 1978 | 73.5 | 38.7 | 112.1 |
| 1979 | 146.6 | 32.9 | 179.5 |
| 1980 | 74.4 | 22.9 | 97.3 |
| 1981 | 65.8 | 7.5 | 73.3 |
| 1982 | 55.2 | 2.9 | 58.1 |
| 1983 | 54.8 | 3.3 | 58.1 |
| 1984 | 46.6 | 1.6 | 48.2 |
| 1985 | 36.6 | 2.6 | 39.2 |

Source: Royal Forestry Department, 1985.

timber. In some places, plantations have been successful but in other regions, they have not. Failure has sometimes been due to factors such as poor adaptation of species to climate and soils, or to outbreaks of pests and diseases. However, failure has frequently occurred because of social problems. Rural population growth, land tenure and distribution, fuelwood collection and the expansion of agriculture are widely recognized as contributing to forest degradation and depletion (Hafner and Apichatvullop, 1990).

The situation is becoming worse in almost all tropical and subtropical countries. As more and more land is needed for agriculture, it will become even more difficult to ensure a sustainable supply of forest products. Increasingly, forest managers will have to take into consideration the needs and problems of the local communities and populations if wood production is to be sustained for future generations.

Designing new strategies to ensure productive and equitable forest systems, while improving the ability to manage both natural and human social systems are among the major issues for those concerned with sustainable forestry (Hafner and Apichatvullop, 1990). Taungya forestry has arisen in response to the need for plantations capable of producing high value timber in regions where local social needs make traditional industrial plantations difficult. Taungya is a system of forest plantations in which peasants plant the seeds or seedlings and tend the trees. As an incentive to participate in taungya, they are allowed to cultivate crops for the first few years between the seedlings of a forest plantation.

Taungya is not synonymous with agroforestry. In agroforestry the focus is upon agricultural crops, and trees are a secondary component. Trees are interplanted so as to complement or supplement production of the agricultural crop(s) selected. In taungya, it is the reverse. Trees and wood production are the primary concern. The distinction is important for this book because there already have been numerous volumes on aspects of agroforestry, such as those by Beets (1982), Huxley (1983), Gomez and Gomez (1983), FAO (1984), Goodland *et al.* (1984), Francis (1986), Altieri (1987), American Society of Agronomy (1988), Nair (1989), Kartasubrata *et al.* (1990), MacDicken and Vergara (1990) and Young (1989).

Trees in agroforestry systems serve many roles such as soil stabilization, production of fodder for animals, supply of mulch for improving soil, improving microclimate, and production of wood for firewood, pulpwood, charcoal and boards for local construction. Commercial production of high quality timber for national or international markets is not usually an objective of agroforestry systems. Rarely does an individual farmer include timber species such as teak within an agroforestry system, because the length of time required for the tree to mature is longer than the farmer's economic horizon. Thus, while agroforestry can solve some of the local wood supply problems, it rarely produces high value timber crops which stimulate the national economy or provide foreign exchange.

The distinction between agroforestry and taungya was a major point that emerged during the symposium from which this book has developed. In that symposium, at the Vth International Congress of Ecology at Yokohama, Japan, it was repeatedly emphasized that while 'agroforestry' covers a broad range of activities, the goal is always to improve agricultural crop production. Tree products are a secondary consideration, and species are chosen to enhance the agricultural production system.

In order to sharpen the focus of the symposium, taungya was also distinguished from other types of 'community forestry' and 'integrated social forestry'. These may start with an established community which is seeking to diversify its production system and increase community self-reliance for wood products. 'Social forestry', 'community forestry', and 'village woodlot forestry', have as a common denominator, a 'social' approach to forestry, that is, an approach that involves local participation in forestry-related activities to meet *local* needs (Kirchhofer and Mercer, 1986). Books dealing with social forestry include the volumes of the ODI (Overseas Development Institute) Social Forestry publications and that of Poffenberger (1990).

Taungya in the sense used in this book generally starts with a governmental agency or private business that wishes to establish or continue a forest products industry, and that recruits families to establish and work the plantation. The workers may be local, or may be imported from other regions. Taungya is an outgrowth of traditional forestry, and has timber production for profit as an ultimate objective. When compared to social forestry, taungya differs in several ways.

**1.** It is concerned with social benefits only as much as such benefits lead to contented and reliable workers.
**2.** The main economic beneficiaries of the taungya are not local peasants, but investors, company or government officials, or segments of the public that stand to benefit from increases in gross national product.
**3.** The role of the local forester is that of 'protector of the forest', in contrast to social forestry systems where he is more of an extension agent at the community level (Kirchhofer and Mercer, 1986).

We have chosen to focus on taungya, as narrowly defined, to emphasize how this approach can contribute to the implementation of sustainable forestry in the tropics and subtropics. We concentrate on taungya as a commercial wood production system that includes crop cultivation as part of the silvicultural scheme. We also include recent developments in taungya, where social benefits for workers are incorporated into the operational scheme. Although there are publications (e.g. Lamb, 1968; Evans, 1982) on many aspects of plantation forestry in the tropics such as selecting the best species for a particular site, there have been few recent publications which deal with taungya as a way for commercial foresters to deal with the social problems arising from increased population growth in rural areas.

For these reasons, we have focused this book on: a description of taungya; a critique of its strengths; an evaluation of its weaknesses; a discussion of ways to overcome the weaknesses. We believe that when economic, social and cultural factors are integrated into the taungya system, it can contribute to the effort to make tropical forestry a sustainable activity. Our objective in this book is to present information that can help forest managers achieve this goal.

While taungya has been tried in Africa (Lowe, 1987; Nwonwu, 1987) and the Americas (Weaver, 1989), it is practised most extensively in Southeast Asia. Recent improvements in the system also have occurred principally in this region, and therefore we have concentrated this book on studies and examples from this region. Part I presents a history and overview of taungya, including its benefits and drawbacks from the social and economic perspective, as well as from the technical (agronomic, forestry) aspects. After having developed an understanding of taungya, we then propose modifications to the system which have the potential to overcome some of the drawbacks, without sacrificing the beneficial aspects. Part II consists of case studies of taungya in Thailand, Indonesia, Nepal, India, the Philippines and China.

## References


Altieri, M.A. (1987) *Agroecology.* Westview Press, Boulder, Colorado.

American Society of Agronomy (1988) *Agroforestry Land-Use Systems.* Nitrogen Fixing Tree Association, Waimanalo, Hawaii.

Beets, W.C. (1982) *Multiple Cropping and Tropical Farming Systems.* Westview Press, Boulder, Colorado.

Evans, J. (1982) *Plantation Forestry in the Tropics.* Clarendon Press, Oxford.

FAO (1984) *Sistemas Agroforestales en America Latina y el Caribe.* Oficina Regional de la FAO para America Latina y el Caribe. Santiago.

Francis, C.A. (1986) *Multiple Cropping Systems.* MacMillan, New York.

Gajaseni, J. and Jordan, C.F. (1990) Decline of teak yield in Northern Thailand: effects of selective logging on forest structure. *Biotropica* 22, 114–118.

Gale, R.P. and Cordray, S.M. (1991) What should forests sustain? Eight answers. *Journal of Forestry* 89(5), 31–36.

Gomez, A.A. and Gomez, K.A. (1983) *Multiple Cropping in the Humid Tropics of Asia.* International Development Research Center, Ottawa, Ontario, Canada.

Goodland, R.J.A., Watson, C. and Ledec, G. (1984) *Environmental Management in Tropical Agriculture.* Westview Press, Boulder, Colorado.

Hafner, J.A. and Apichatvullop, Y. (1990) Farming the forest: managing people and trees in reserved forests in Thailand. *Geoforum* 21, 331–346.

Huxley, P.A. (1983) *Plant Research and Agroforestry.* International Council for Research in Agroforestry. Nairobi, Kenya.

Kartasubrata, J., Tjitrosomo, S.S. and Umaly, R.C. (eds) (1990) *Symposium on Agroforestry Systems and Technologies, Bogor, Indonesia, 19–21 Sept., 1989.* BIOTROP Special Publication No. 39. SEAMEO BIOTROP, Box 17, Bogor, Indonesia.

Kirchhofer, J. and Mercer, E. (1986) Putting social and community forestry in perspective in the Asia–Pacific region. In Fujisaka, S., Sajise, P.E. and del Castillo, R.A. (eds) *Man, Agriculture and the Tropical Forest: Change and Development in the Philippine Uplands.* Winrock International Institute for Agricultural Development. Bangkok, Thailand, pp. 323–335.

Lamb, A.F.A. (1968) Artificial regeneration within the humid lowland tropical forest. *Unasylva* 22(4), 7–15.

Lowe, R.G. (1987) Development of taungya in Nigeria. In Gholz, H.L. (ed.) *Agroforestry: Realities, Possibilities and Potentials.* Martinus Nijhoff, Dordrecht, pp. 137–154.

MacDicken, K.G. and Vergara, N.T. (1990) *Agroforestry: Classification and Management.* Wiley, New York.

Nair, P.K.R. (1989) *Agroforestry Systems in the Tropics.* Kluwer, Dordrecht.

Nwonwu, F.O.C. (1987) Cost minimization through the use of taungya in pulpwood plantation establishment. *Agroforestry Systems* 5, 455–462.

Poffenberger, M. (ed.) (1990) *Keepers of the Forest. Land Management Alternatives in Southeast Asia.* Kumarian Press, West Hartford, Connecticut.

Royal Forestry Dept (1985) *Forestry Statistics of Thailand, 1985.* Royal Forest Department of Thailand, Bangkok.

Weaver, P.L. (1989) Taungya plantings in Puerto Rico. *Journal of Forestry* 87 (3), 37–41.

Young, A. (1989) *Agroforestry for Soil Conservation.* CAB International, Wallingford, UK.

# PART I
## The Taungya System

# Chapter 1
# Overview of Taungya

Jiragorn Gajaseni

*Department of Biology, Faculty of Science,*
*Chulalongkorn University, Bangkok 10500, Thailand*

Like most other developing regions, Southeast Asia has experienced a rapid transformation of its economic and social structure during the last two decades. However compared with fast growing sectors such as industries and services, forestry and agriculture, traditional elements of the national economy, have made relatively slow progress. Many lowland floodplains already are under cultivation and there is little room for expansion. Most of the accessible forests have been exploited and their regeneration in the near future is doubtful. Meanwhile as population grows, the pressure for agricultural development continues to increase.

Despite these problems, the largest opportunities for economic expansion are in forestry and agriculture. However to realize economic gains, effective planning is becoming increasingly necessary. Continued production and growth will depend on increased intensification and diversification. Forestry and agricultural policies must be well coordinated. Meanwhile, increasing environmental and social concerns limit options in natural, managed natural, and manmade ecosystems. There is a basic necessity to understand social and ecological processes in those systems, in order to devise management strategies that will result in sustainability of production and environmental amelioration.

## Historical Perspective

Evidence from excavations at Ban Chieng, Northeastern Thailand, indicate that people have cultivated plants in the area for centuries (Gorman, 1969). Systems involved combinations of trees, root crops, and seed crops. They are thought to be the origin of regional agriculture (Gorman, 1971).

Primitive hunting, fishing and gathering was also carried out in the area.

The earliest important crops of Southeast Asia may have been root crops such as taro (*Colocasia esculenta*) and yam (*Dioscorea alata*), both of which preceded rice (*Oryza sativa*). Other crops of tremendous importance to the cultivators were perennials such as banana (*Musa* spp.), papaya (*Carica papaya*), coconut palm (*Cocos nucifera*), jackfruit (*Artocarpus* spp.), mango (*Mangifera indica*) and durian (*Durio zibethinus*) (Pelzer, 1978).

Archaeological evidence from Thailand indicates that it was one of the first rice growing regions of the world (Sud Saengwichien, 1974). There have been two major varieties of rice. Upland rice was planted in dry, mountainous areas. Swampy rice was used in lowland, alluvial floodplains. Rice has been used in both upland shifting cultivation, and in lowland permanent paddy rice cultivation (Gorman, 1971).

Shifting cultivation may be the oldest system devised by humans for use of forest ecosystems. Southeast Asian people have practised unirrigated farming for centuries, using methods variously referred to as slash and burn, swidden agriculture, or shifting cultivation. These practices, often thought of as primitive, are based on an intrinsic ecological rationale, including manipulation of the cultivation – fallow cycle.

Shifting cultivation does not necessarily imply that the cultivators themselves are nomadic. Some of them live in villages that have been settled for hundreds of years. The cultivators have made repeated use of the same fields. Some farmers combine or supplement their use of permanent irrigated fields with shifting cultivation, especially where land is scarce.

Many shifting cultivation societies have developed balanced systems which are able to persist in the same locality for many generations (Kunstadter *et al.*, 1978). Nevertheless, the number of people that can be supported by shifting cultivation is limited. In recent years, increases in population, decline in the availability of land, psychological ties to the market economy and increased cost of purchased items necessary for agricultural products have upset what had been a relatively self-sustaining system.

The number of people involved in shifting cultivation in Thailand was estimated to have risen from 300000 in 1967 to over 700000 in 1982 (Kunstadter and Chapman, 1978). In many Southeast Asian countries, sufficient fallow often is not given, because of social, political or economic conditions beyond the control of the farmer or communities. As a result, shifting cultivation is no longer sustainable in these regions, and this could threaten future productivity of the land.

## Taungya Systems

Taungya is a land management system with origins in Southeast Asia. In the sense that the term applies to the land management system used in Burma (Myanmar) by the British Colonial foresters, taungya is a reforestation system which combines agricultural crops with forest tree seedlings. Peasants were allowed to cultivate crops for the first few years amongst the seedlings of teak plantations. There were no social development schemes, and no wages were paid to the peasants for their labour. Since the beginning of the 20th century the Royal Forestry Department of Thailand has promoted the taungya system of forest plantations with agriculture, to increase the production of teak (*Tectona grandis*) and to mitigate the problems caused by shifting cultivators in native forests (Banijbatana, 1978).

In 1967, the Forest Industry Organization of Thailand began the 'Forest Village System', a modification of the taungya system. In the Forest Village System, peasants are recruited to form villages with approximately 100 families in each unit. Each household is allotted 1 rai (0.16 ha) of land for permanent settlement and home gardens. Health care, education, and other social services are provided for the families. Their obligation is to clear and plant 10 rais of teak per year. Each plantation unit is assigned to reforest 1000 rais annually.

Seeds for teak are collected in special seed orchards. The seeds are cultivated in beds for one year. Then 'stumps' are prepared by pruning the seedlings and leaving about 2 cm of stem and 12–22 cm of root. The stumps are planted by making a hole with a stick, inserting the stump in the hole, and tamping the soil around the stump. The first plantations were planted with stump spacings of 3 × 3 m. Now, stumps are planted at 4 × 4 m intervals. Plantations were established with the expectation of a 60-year rotation.

Plantation workers are allowed to cultivate annual or biennial crops of their choice between the teak trees for 2 years after the trees have been planted. However, they often cultivate a garden for only the first year. It is difficult to continue cultivating a garden the second year, because it involves continually moving back and forth between the new teak plantation and the old garden, just at the time of peak labour requirements.

In the traditional taungya method, these subsistence crops were the only benefit for the peasants. In the Forest Village System, the workers are paid for clearing, planting, and weeding the site. In 1991, the daily rate was 45 baht for men (US$1 = 25.5 baht) and 42 baht for women. Other family members can also do contract work for wages by doing additional weeding or thinning on older plots. If, after 3 years in the system, a family has cleared and planted 30 rais, the Forest Industry gives a substantial bonus, depending on the survival rate of seedlings in the plot under their custody.

After longer periods of membership in the village and fulfilling obligations, 5 rais will be allotted to each family member for permanent cultivation (FIO, 1977).

## Taungya in Other Regions

In Indonesia, the Forestry Estate Enterprise has used taungya for reforestation, especially teak (Siregar, 1990). There the system is called 'Tumpangsari'. Teak is planted in areas with a pronounced dry season, and the expectation is for an 80-year rotation. Most plantations have been established in areas with rather poor soils and slopes up to 30%.

In Tumpangsari, a contract is made between individual farmers and the Forestry Estate Enterprise. The contract specifies the area to be planted, the period of agricultural cultivation, and crops to be planted, and the amount of money to be paid for tree planting and tending. A 0.25 ha plot is allotted per family. Intercropping is allowed for 2 years, and crops include dryland rice, maize, pepper, peanuts, and soybeans. Under certain conditions, cassava also is permitted. On sloping land, erosion control structures are built with stones or 'trash' (dead trees and branches). Other erosion control measures include drainage ditches, and planting along contours (Wiersum, 1982).

In Peninsular Malaysia and Sabah, subsistence crops including bananas, maize, and upland rice are planted between rows of teak, gmelina (*Gmelina arborea*) and several species of pine. Recently, trials have begun with cacao (*Theobroma cacao*) and rattan (*Calamus* sp.) between rows of rubber trees (*Hevea brasiliensis*). However, taungya in Malaysia is still insignificant in terms of area and economics, compared to even-aged monoculture plantations of oil palm, rubber, and pepper (*Piper nigrum*) (Sani, 1990).

In the Philippines, the taungya system has not been undertaken to any large extent. There has been some success with *Pinus kesiya* in efforts sponsored by the government (Lasco and Lasco, 1990). Other species that have been established experimentally using taungya are *Eucalyptus deglupta* and *Pittosporum resiniferum.*

In Vietnam, indigenous tree species including *Manglietia glauca, Cinnamomum album, Styrax tonkinensis,* and *Dipterocarpus alatus* have been found to be particularly suited for taungya (Cuc, 1988). Mixed species plantations also are being tried in Vietnam, including *Eucalyptus camaldulensis* interplanted with fast growing legumes such as *Acacia auriculiformis* and *Sesbania grandifolia,* and *Hopea odorata* interplanted with *Cassia siamea* (Cuc, 1988).

## Acceptance of Taungya

Surrounding traditional rural villages there have usually been fields that were cultivated in a variety of systems which reflect the natural and social environment. Many cultivators imitated natural systems by planting a mixture of root crops, herbs, climbers, shrubs, trees, and grain crops such as maize and upland rice (Harris, 1972). Thus, from an ecological and agricultural perspective, conditions in Thailand and other regions of Southeast Asia have been appropriate for adoption of taungya. Also, taungya appeals to the desire of peasant farmers for permanent settlement.

The major problem with the system as it exists in most regions is that under taungya as a system of forestry, the land remains under the ownership of the government or commercial enterprises. The farmer's desire for outright land ownership is often frustrated. In Southeast Asia a usual first step toward achieving these goals is to plant trees. This gives land rights and land tenure in the traditional settlement pattern. However, in the taungya system it does not.

A second difficulty in the taungya system is that the long rotation time means that after the initial year or two of crop cultivation, the land does not produce any income until the final harvest, which may not occur for 30–60 more years.

## Conclusion

Taungya is a system of plantation forestry in which crops are planted between the tree seedlings, for the first few years of the rotation. In the forest village system of taungya, peasants participating in the system are given additional benefits.

Taungya solves two major problems: it provides for re-establishment of high value timber that will provide much needed income in future years; it stabilizes the lives of peasants, and reduces the destruction caused by shifting cultivation.

Taungya systems generally face two major problems: peasant farmers generally wish to own their own land, and this is not possible under taungya systems as they are now operated; there is an extremely long (by economic standards) interval between investment in a plantation and income from that plantation.

These problems and benefits will be examined more closely in the following chapters.

## References

Banijbatana, D. (1978) Forestry policy in northern Thailand. In Kundstadter, P. Chapman, E.C. and Sabhasri, S. (eds) *Farmers in the Forest.* University of Hawaii Press, Honolulu, pp. 54–60.

Cuc, L.T. (1988) *Agroforestry Practices in Vietnam.* Environment and Policy Institute, East-West Center, Working Paper No. 9. Honolulu, Hawaii.

FIO (1977) *Criteria and Obligation to be a Member of the Forest Village System.* Forest Industry Organization of Thailand (in Thai).

Gorman, C.F. (1969) Haobinhian: pebble-tool complex with early plant association in Southeast Asia. *Science* 163, 671–673.

Gorman, C.F. (1971) The Haobinhian and after: subsistence patterns in Southeast Asia during the Late Pleistocene and Early Recent Period. *World Archeology* 2, 217–240.

Harris, D.R. (1972) The origin of agriculture in the tropics. *American Scientist* 60, 180–193.

Kunstadter, P. and Chapman, E.C. (1978) Problems of shifting cultivation and economic development in Northern Thailand. In Kunstadter, P., Chapman, E.C. and Sabharsi, S. (eds) *Farmers in the Forest,* University of Hawaii Press, Honolulu.

Kunstadter, P., Chapman, E.C. and Sabharsi, S. (eds) (1978) *Farmers in the Forest.* University of Hawaii Press, Honolulu.

Lasco, R.D. and Lasco, H.D. (1990) Agroforestry systems in the Philippines: Inventory and assessment. In *Agroforestry Systems and Technologies,* BIOTROP Spec. Publ. No. 39. South-East Asian Regional Centre for Tropical Biology. Bogor, Indonesia, pp. 45–56.

Pelzer, K.J. (1978) Swidden cultivation in Southeast Asia: historical, ecological, and economic perspectives. In Kunstadter, P., Chapman, E.C. and Sabharsi, S. (eds) *Farmers in the Forest,* University of Hawaii Press, Honolulu, pp. 271–285.

Sani, H.B. (1990) Agroforestry in Sarawak, Malaysia: recent developments. In *Agroforestry Systems and Technologies,* BIOTROP Spec. Publ. No. 39. South-East Asian Regional Centre for Tropical Biology. Bogor, Indonesia.

Siregar, A.M. (1990) Implementation of agroforestry in Indonesia. In *Agroforestry Systems and Technologies,* BIOTROP Spec. Publ. No. 39. South-East Asian Regional Centre for Tropical Biology. Bogor, Indonesia, pp. 29–36.

Sud Saengwichien (1974) *The Pre-history of Thailand.* Mahidol University Press.

Wiersum, K.F. (1982) Tree gardening and taungya on Java: examples of agroforestry techniques in the humid tropics. *Agroforestry Systems* 1, 53–70.

# Chapter 2
# **Origins of Taungya**

Shinya Takeda

*Laboratory of Forest Resources, Faculty of Agriculture, Kyoto University, Kyoto 606, Japan*

It is sometimes claimed that taungya originated in Burma in the mid-1800s (Evans, 1982). However, there are undocumented stories of its use in much earlier times. Professor Pan Weichou, former Vice President of Central-South Forestry University, Zhuzhou, Hunan, Peoples Republic of China, spent much of his career in studying plantations of Chinese fir (*Cunninghamia lanceolata*) in western Hunan. He claimed that the nomadic tribes which are locally known as 'over the Hill People' had the custom of planting crops together with Chinese fir, harvesting the crops, selling or abandoning the land, and moving on to a new site. Today in Northern Thailand, Burma and Laos, there are hill tribes called the Miao or Hmong and the Yao. These people are descended from the 'over the hill' tribes in southwestern and south-central China, and migrations may have taken place for thousands of years (Matisoff, 1983).

A type of agroforestry system called 'shamba' which is similar to taungya has been popular in agricultural areas of Kenya since the early 1900s (Oduol, 1989). The main difference between this system and taungya in Burma is the considerable integration of the cultivators into the Forest Department (civil service status) in Kenya. Similar systems arose later in Nigeria (Lowe, 1987) and Trinidad (Evans, 1982).

Although it is difficult to attribute a specific time and place to its origin, it is clear that the term 'Taungya', as used for a reforestation technique, originated in Burma. It is the Burmese term for temporary cultivation on hill land. *Taung* means hill, and *ya* means cultivation as practised in the hills of Burma. In this system, trees are combined with agricultural crops in cultivated fields. Taungya for reforestation purposes was an adaptation of traditional shifting cultivation. In taungya for reforestation, swidden cultivators were allowed to clear forest and grow crops such as sweet potatoes, cotton, and chillis on the condition that they also planted teak or other valuable species (Blanford, 1925).

## Taungya in Burma

The history of forestry in Burma is relatively well documented. The accounts give a picture of the conditions there that gave rise to the system of taungya.

In 1827, a Dr Wallich was sent to report on the natural resources of Tenasserim Province, at that time the only part of Burma under British rule (Blanford, 1958). He was so alarmed (by the overcutting) that he recommended strict government control. Instead, the forests were opened to speculation, largely to raise revenue for the government as quickly as possible (Stebbing, 1922).

In 1833, a headman and 10 villagers were employed under orders of the Civil Commissioner Maingy to collect and plant seedlings of teak. This was the modest beginning of a Forest Department in Burma (MacGillivray, 1990). Three years later, Lord Auckland deputed another medicobotanist, Helfer, to examine the Tenasserim forests. The report discusses girdling, disease, fire, Taungya, and the lack of natural regeneration, and recommended plantations (Blanford, 1958).

In 1841, Wallich submitted a further report in which he listed reasons for the failure of natural regeneration of teak: fire, browsers and competitive re-growth of other species. Clearly, there was already an appreciation that silviculture would be necessary.

Most accounts of taungya credit Dietrich Brandis with the origination of the system in Burma. He was the forest officer in charge of the teak forest of Pegu from 1856 to 1862, and was appointed the first Inspector-General of Forest in British India in 1864. Brandis had been a German forester, and was familiar with the 'Waldfeldbau' system, a combined system of field crops and high forest, and with 'Hackwaldbetrieb' or ('Haubergswirtschaft'), a combined system of field crops and coppice. Thus it seems likely that Brandis, knowing the efficiency of taungya type agroforestry in Germany, introduced the system and applied it to establish plantation forests in Burma. One of the first taungya was made by a Karen tribesman, U PanHee, in the Tharrawaddy Division, as a personal present to Dietrich Brandis in 1856 (Blanford, 1958). Brandis remarked 'This, if the people can ever be brought to do it, is likely to become the most efficient mode of planting teak in this country' (Blanford, 1958).

However, the credit of applying taungya in its practical details in the Tharrawaddy forests appears to belong to Colonel W.J. Seaton, who was in charge of those forests from about 1863 to 1870. Annual Reports on Forest Administration in Burma show records of teak plantations beginning in 1863 (MacGillivray, 1990). Taungya plantations were commenced in Tharrawaddy on a small scale in 1869, and since 1873 have been extended almost continuously. These plantations have proved a cheap and efficient means of raising teak plantations on a large scale, and are admirably adapted to the habits of the jungle population (Troup, 1921, p. 733). In other reports Troup (1928, p. 11; 1940, p. 174) said that taungya was started in 1869.

Rate of plantation establishment peaked in the early 1890s, at about 5000 acres per year. Teak grown in plantations was susceptible to outbreaks of pests such as the bee-hole borer (*Xyleutes ceramicus*) and the defoliators and leaf-skeletonizers (*Pyrausta machaeralis, Eutectona machaeralis, Hyblaea puera*, and *Hapalia machaeralis*) (FAR, 1902). However, the greatest constraint that prevented the Forestry Department from achieving their goal of establishing 3000 acres per year in plantation was labour. The preferred method of obtaining labour was through taungya.

By the late 1890s, taungya plantations began to decline, and fell to a rate of less than 1000 acres per year in 1909. Taungya plantations had been scattered in small blocks throughout the natural teak forests, and it was found that organized tending and thinning, necessary for survival of the trees, were almost impossible. The reason that the plantations were scattered was the reluctance to establish taungya in areas in which there was already native teak forest. As a result of selection of sites on which there was no native teak forest, many plantations were made on soil which was not suitable for teak (Blanford, 1925, p. 82).

After 1906, taungya plantations were almost given up in most of Burma, and attention was turned to the improvement of natural stocks by means of improvement fellings. The result was that the intensity of felling increased. Eventually, the felling came near to a regeneration felling and in some cases was actually a seeding felling. The cost of such felling was high, and the result of natural regeneration was extremely uncertain. Therefore, the taungya system was re-introduced in 1918 for the Tharrawaddy Yoma forest and then other parts of Burma (Blanford, 1925).

The problem of obtaining labour for the plantations still remained. W.J. Keith, Revenue Secretary in the Government of Burma Proceedings (FAR, 1917) stated:

> The difficulty has been to discover adequate and suitable measures for coping with the situation. The wild tribes which practice this method of cultivation are comparatively low in the scale of civilization and are averse to settling in the plains. Efforts have been made to induce them to undertake permanent cultivation, but without success, and for political reasons it has been impossible to bring pressure to bear on them to abandon their nomadic habits. A

> promising solution to the problem lies in the experiment now being made of using them to re-afforest the areas which they destroy. The difficulties are many and obvious but the advantages to be gained are overwhelming. Taungya cultivation uncontrolled causes greater and more lasting damage than a fire. The same system if successfully controlled should provide the jungle tribes with all the virgin soil they require and at the same time secure that the taungyas when abandoned are stocked with a valuable forest crop. Shortage of labour has always been the chief obstacle in forest development in Burma, and this experiment, which aims at regulating taungyas by rotation, is calculated to serve an adequate and suitable supply of labour and also to create feelings of confidence and amity in the minds of the people towards the Forest Officers whose activities in matters of reservation have hitherto not unnaturally been viewed with dislike and suspicion. The problem affects forest development as a whole and is at present under consideration in connection with the control of forest villages.

Plantations using taungya increased rapidly again after the First World War. The working plan of Blanford, Divisional Officer of Tharrawaddy, required clear felling and intensive silviculture, and depended on plentiful labour. Revenue Secretary Lewisohn summarized the situation since Blanford had first questioned Brandis' system in 1914:

> The past five years have witnessed a growing interest in silvicultural problems. Whereas previously the mixed and uneven-aged character of the forests was held to make it impossible to break away from the diffused method of working known as the 'Selection System', it is today generally recognized that it need no longer be confined to small areas in a few specially favourable localities. Artificial regeneration necessarily implies concentration of effort, and the remarkable success of plantations created through the agency of taungya cutters points strongly to the advantages of clear fellings from a silvicultural point of view. The size of the area clear-felled in any one place need simply be limited by the amount of ground that can be covered with young forest growth by artificial means. This in itself is a great step forward in the development of the forest in the Province. It makes it possible to make profitable use of mechanical methods of extraction of timber.
>
> *FAR, 1917, 1918*

Artificial regeneration by plantations gained in favour. Leete stated:

> To a man, all of us now want to discard the 'selection system' with its reliance on natural reproduction in favor of more or less 'clear fellings' with artificial regeneration, par excellence by taungya. If an

> average forest now worked by selection fellings were managed intensively with clear fellings and concentrated artificial regeneration of even aged crops the increment could undoubtedly be multiplied, anything from ten to twenty fold. Actual plantations up to fifty years old support this view. It is something big for Foresters to work for.
>
> *FAR, 1919*

By 1923, Leete's enthusiasm for artificial regeneration was damped by the economic depression. The change of heart is illustrated in FAR reports (1919–1920, 1923–1924):

> Plantation work is of absorbing interest and has proved to be of steadily increasing importance both for the general enrichment of the forests and for the special needs of the local population in the way of house building timber and fuel. The fact however remains that within the present generation there is no likelihood of plantations being established over anything more than a very small fraction of the reserved forests of the province. It may indeed come as a surprise to some people to be told that anything appreciable can be done to improve and accelerate forest growth other than by plantations.

This cautious retraction was accelerated the following year. By 1924, Leete was openly admitting:

> We were infected by the world-wide optimism which marked the close of the Great War ... subsequent trade depression has done much to disillusion us. There was an overswing of the pendulum in favour of concentrated methods to which the admirable Working Plan for the Tharrawaddy Yoma forests contributed very considerably. There were many admirers and imitators of this Working Plan, and in many cases the latter were responsible for an ill-considered and unsympathetic rush at concentrated regeneration without proper consideration for the waste involved.

## Taungya Since 1920

The programme of work in the system of taungya practised since the 1920s in Burma is outlined in Table 2.1. In early April, the fields were burned. In May, unburned refuse was heaped up and burned again. Then small bamboo stakes were used to mark the spacing where seeds were to be planted. The spacing was generally 6 feet by 6 feet. After the tree seeds were planted, crops were sown. Field crops were harvested in January. Lines were cleared and counts were made of surviving seedlings. Then payment was made to the farmers.

**Table 2.1.** A typical programme of work to be carried out in the taungya system in 1920 in Burma.

| | |
|---|---|
| *The year before regeneration* | |
| December | Allotment of areas to taungya participants to be completed. |
| January | Cutting to commence. |
| March | Cutting to be completed. |
| *First year of regeneration* | |
| March and April | Collection of seed to be completed. |
| April | Burning to be completed not later than 20th April. |
| May | Heaping and reburning of unburnt refuse to be completed. |
| May | Staking and sowing of tree seeds at stake and broad casting of accessory species to be completed. (In the case of teak, sowing should if possible be completed on suitable plots throughout the area by 30th April.) |
| May to June | Sowing of field crops. |
| July | Inspection of lines and stakes and planting up of blanks. |
| July to November | Tending of tree seedlings in conjunction with the field crops. |
| November and December | Reaping of field crops. |
| January | Clearing of lines, restaking and counting of seedlings surviving and payment of taungya participants. Survey to be carried out and inspection paths made. |
| March | Fire protection. |
| *The second year* | |
| April and May | Fire protection to be continued up to the break of the rains. |
| May | First weeding and filling of gaps to be completed by the end of May or 15th June at latest. |
| September | Second weeding. |
| March | Fire protection. |
| *The third year* | |
| April and May | Fire protection until the break of the rains. |
| June | First weeding. |
| September | Second weeding if necessary. |
| March | Fire protection. |
| *The fourth year* | |
| April and May | Fire protection until the break of the rains. |
| June and July | Cleaning if necessary. |
| March | Fire protection. |
| *The fifth year* | |
| April and May | Fire protection. |
| November and December | Cleaning, creeper cutting and thinning. |

After Blanford, 1925.
Note: The year should be taken as the Forest year, i.e. 1st April to 31st March.

In lower Burma, the farmers cultivated an area for only one year, and then shifted to another area. In upper Burma, sometimes it was possible to cultivate crops in a plantation for a second year. This sequence of work is very similar to that currently practised in the 'Modified Taungya System' (Forest Village System) used today in Thailand, and described here in Chapter 8. The Forest Villages of Thailand, thought to be unique to the Modified Taungya System there, had already been introduced into Burma during the 1920s.

In the Forest Villages,

> The villagers are given free building materials, free land for taungya cultivation and are sometimes let off all tax. In addition, the head man usually receives some regular payment and the taungya cutters are paid by the results of the regeneration in their taungya. Advances for purchase of food, seed, rice, etc. were given to them when they first enter a village, or when they have bad crops. When they are free from the taungya work they are employed on forest work for which they get the ordinary rates of pay.
>
> *Blanford, 1925, p. 93*

The only difference between this system in Burma of the 1920s and the modern Forest Village System in Thailand is the method of planting. In Burma, seeds were sowed directly. In Thailand, seeds are planted in nursery beds, and when seedlings develop, 'stumps' are cut from the stem and root, and these 'stumps' are planted in the taungya sites.

Teak plantations came to a halt during the Second World War. After the war, commercial plantations were begun. The major planting programme was launched in 1980 'with a starting target of 40000 acres per year, to reach a final target of 90000 acres annually and thereafter' (Government of Burma, 1989). This was a highly optimistic target, given the British record, and interestingly, it seems to be running into familiar problems (MacGillivray, 1990): 'the programme was so heavy to cope with and there are difficulties getting labour for cultural works to be done in time' (Government of Burma, 1989).

## Conclusion

The history of forestry is replete with debates on the relative merits of selective cutting with natural regeneration vs. clearcutting followed by artificial regeneration in plantations. Each approach has many factors to consider. When shifting cultivators inhabit forest reserves, they constitute a problem for natural regeneration. Incorporating them into a taungya system of artificial regeneration may solve that problem to some extent, but there are other social and economic problems associated with taungya. One problem is the reluctance of nomadic tribes to give up their traditional way of life. From the forester's point of view, their reluctance to work is a problem. 'Most ... seem to prefer to have six days of rest and the seventh of labour' (FAR, 1917).

Taungya in modern Southeast Asia is changing. The preference for a nomadic lifestyle no longer seems to be a major obstacle, although it still may be a local problem. Other concerns such as land tenure and economic factors have become predominant in taungya. These are discussed in the following chapter.

## Acknowledgements

Many original documents relevant to the history of taungya are not available to us. A.W. MacGillivray cited some of them in his report *Forest Use and Conflict in Burma, 1750–1990* submitted to the Imperial College of Science, Technology and Medicine (University of London). Where material from original documents was taken from MacGillivray's report, the citation is so noted. We thank Mr MacGillivray for supplying us with a copy of his report.

## References

Blanford, H.R. (1925) Regeneration with the assistance of taungya in Burma. *Indian Forest Records* 11(III), 81–121.

Blanford, H.R. (1958) Highlights of one hundred years of forestry in Burma. *Empire Forestry Review* 37(1), 33–42.

Evans, J. (1982) *Plantation Forestry in the Tropics.* Clarendon Press, Oxford.

FAR Reports. These are Annual Reports on Forest Administration in Burma, 1861–1940. The citations given here are taken from MacGillivray (1990), who states that reports for most years are in the India Office Records and Libraries.

Government of Burma (1989) *Forestry Situation in Myanmar.* Forest Department, Rangoon. Cited in MacGillivray (1990).

Lowe, R.G. (1987) Development of taungya in Nigeria. In Gholz, H.L. (ed.) *Agroforestry: Realities, Possibilities and Potentials.* Martinus Nijhoff, Dordrecht, pp. 137–154.

MacGillivray, A.W. (1990) *Forest Use and Conflict in Burma: 1750–1990.* Imperial College of Science and Technology and Medicine, University of London. Report submitted in partial fulfillment of the requirements for the MSc degree and/or the DIC.

Matisoff, J.A. (1983) Linguistic diversity and language contact. In McKinnon, J. and Bhruksasri, W. (eds) *Highlanders of Thailand.* Oxford University Press.

Oduol, P. (1989) The shamba system: an indigenous system of food production from forest areas in Kenya. In: Nair P.K.R. (ed.) *Agroforestry Systems in the Tropics.* Kluwer Academic Publishers, Dordrecht, pp. 401–410.

Stebbing, E.P. (1922) *The Forests of India*, Vol. 1. Bodley Head, London. Cited in MacGillivray (1990).

Troup, R.S. (1921) *The Silviculture of Indian Trees*, Vol II. Oxford University Press, London, pp. 739–741.

Troup, R.S. (1928) *Silvicultural Systems*. Oxford University Press, London.

Troup, R.S. (1940) *Colonial Forest Administration*. Oxford University Press, London, 476 pp.

# Chapter 3
# Socioeconomic Aspects of Taungya

Jiragorn Gajaseni

*Department of Biology, Faculty of Science, Chulalongkorn University, Bangkok 10500, Thailand*

Taungya has often been praised as a panacea to solve problems of shifting cultivation and reforestation in the tropics. Theoretically, the system should be highly desirable, because it represents a step toward solving these major problems in less-developed countries. Despite the supposed advantages of the system however, taungya has not been widely adopted. In this chapter, the idea is put forward that despite the benefits of the system from an agronomic and forest management point of view, the system has not realized its potential because of several inherent limitations. These limitations are social and economic in nature.

Taungya, in the sense that the term applies to the land management system used in Burma by the British Colonial foresters (see Chapter 2), is a reforestation system which combines agricultural crops with forest tree seedlings. Peasants were allowed to cultivate crops for several years amongst the seedlings of teak plantations. There were no social development schemes, and no wages were paid to the peasants for labour.

Development schemes based purely on technical grounds and national economic criteria are doomed to failure if the local people, and their social and economic problems are not taken into consideration. These latter factors are essential elements of planning for successful development (FAO, 1984). An emphasis on the social aspects of tropical forest resource management has recently been emphasized as the key to success or failure of management schemes (Ashton, 1977; Eighth World Forestry Congress, 1978; Budowski, 1982; FAO, 1984; Vergara *et al.*, 1986).

In the following sections, the social and economic problems associated with taungya are presented, and in the final sections, some solutions to the problems are discussed. However, in order to gain an appreciation of how the current difficulties arose, it is important to discuss first the history of taungya from a social and economic viewpoint.

## Historical Perspectives

As mentioned above, taungya in the sense used by the British Colonial Foresters, is merely a land management system. At the time that Burma (Myanmar) was colonized and known as British India, the system of taungya in which peasants were allowed to plant crops during the first few years of a teak plantation was well known. It was implemented, in part, as a means of providing cheap labour for teak reforestation, and to reduce deforestation due to shifting cultivation or other reasons. It was also implemented as a means to control local populations and to prevent uprisings against the colonial government. Peasants were dispersed to outlying locations, thereby making their society easier to control.

There was little in the taungya reforestation system of Burma that was related to the socioeconomic improvement of the underprivileged rural people. Taungya in the colonial days was simply an exploitation of land and people, and was used as a political element to establish authority.

The development of taungya in Indonesia was likewise marked in early years by colonial exploitation (Wiersum, 1982). Taungya systems similar to that in Burma and Indonesia were also developed in Japan and Thailand, even though these countries were not colonized. The system practised in Japan was known by the local name of Yakihato Ringyo (Ariki, 1986). The system appeared at the same time as the advent of capitalism in Japan, which brought about large-scale forest ownership by feudal lords. Arable land became scarce, and rural peasants who had lost their own land often entered into an agreement with the feudal lords. The usual arrangement was that the peasants had to afforest the land after using it for cultivation. The system disappeared in Japan, when opportunities arose for work in factories during the era of industrialization. Even in rural areas, the lower classes could earn sufficient wages from industrial employment to secure daily necessities, and consequently, they abandoned the practice of Yakihato Ringyo.

The taungya reforestation system was begun in northern Thailand around 1906 (Banijbatana, 1978). All land belonged to the King, and was allotted to citizens according to their social status. However, lowland and hill tribes were not included, even though they had occupied the land for generations. As a result, the hill tribes of northern Thailand that were not part of the socioeconomic sphere of the national economy, faced severe problems. The hill tribes traditionally maintained themselves through shifting cultivation, and nationalization deprived them of their source of crops. As a result, taungya was instituted in northern Thailand as part of the sociopolitical aspects of the national development plan.

This brief overview suggests that there are certain socioeconomic conditions which give rise to taungya, and that development of the system is independent, to a certain extent, of the population density, amount of arable and forested land, and degree of development. While agricultural and forest management practices are determined primarily by local physical circumstances and ecological characteristics, the choice of any development system is determined by socioeconomic factors (Wiersum, 1982).

Regions where taungya has developed have certain socioeconomic characteristics in common.

**1.** Large economic gains can be attained through illegal harvesting of indigenous timbers, and consequently there is a need to control the harvested lands.
**2.** There is a need for reforestation activity.
**3.** There is a large human resource with need for subsistence and employment.

The taungya system is an arrangement between the legal landowner, which may be the state, a private landlord, or a business enterprise, and the peasant who lives on the land and is the *de facto* manager. The system has evolved where there is a shortage of land available for peasants to cultivate, and where they have no other suitable means for alleviating their poverty.

Despite the relief that taungya has provided to some rural inhabitants, the major benefit of the system has accrued to the legal landowners. In many cases, the local peasants still suffer hardships due to socioeconomic difficulties with the taungya system. The following sections describe problems faced by peasants that have not been satisfactorily solved by the taungya system as described above.

## Problems with 'Colonial' Taungya

### Lack of permanent settlement

Although cultivators in tropical countries frequently are referred to as 'shifting cultivators', most of them are in fact not at all nomadic. Some of them live in villages, with their houses surrounded by a diversified 'home garden', in a tradition that is hundreds of years old (Kunstadter and Chapman, 1978). Home gardens have an important significance in Thailand, Kalimantan, Indonesia, and other regions of the humid tropics (Siebert, 1990). Planting of a tree is the first step in a settlement pattern, and confers land rights and land tenure. Surrounding the village may be a series of fields that are cultivated cyclically, in a systematic pattern of reuse and reforestation of the same parcels of land.

Owning a piece of land for agriculture, and achieving permanent settlement is an ultimate goal of most peasants in an agriculturally-based society. Not only does it provide them income, they can also grow plants that give them subsistence, medicines, fuel, and other necessities. Owning land gives the cultivator an intimate relationship with the land.

Taungya does not confer the benefit of land ownership to the peasant farmer. Its sole benefit to him is that it allows the farmer access to agricultural land for which he has no title. He does not gain from carrying out good stewardship of the land. Taking good care of seedlings can actually be counterproductive for the peasant farmer in a taungya reforestation project. If he tends and cultivates the trees conscientiously and efficiently, the trees will grow vigorously, and he may be forced to move all the sooner to a new area opened for planting. He will be forced to leave his residence, and find or establish a new one. Thus, the farmer is essentially punished for good stewardship, and rewarded for indifference in tending the trees.

Taungya as practised in the past will never achieve a goal of permanent settlement for rural land development. Taungya will remain merely 'swidden afforestation', a simultaneous cultivation of food crops and trees at early stages, and a later evolution to pure forest.

Another problem with taungya is that rural life is sometimes perceived as dull and uninteresting, compared to that in urban areas. This perception increases the tendency of young taungya cultivators to migrate to urban centres where life is less laborious and monotonous, and where they believe there may be better employment opportunities. This situation was recognized as a serious threat to taungya in Nigeria (Enabor *et al.*, 1982).

## Employment and under-employment

Another problem commonly faced by tropical agricultural societies is the need to provide employment for a rapidly increasing population in rural areas. The problem is mainly under-employment, a condition in which irregular work of low productivity results in extremely low levels of earnings and consumption (World Bank, 1982). Under-employment accelerates seasonal movement from the countryside to jobs in urban areas during the non-growing season. This migration increases urban problems, because it is difficult to reduce urban unemployment if under-employment and low earnings force rural workers to seasonally migrate to the cities (World Bank, 1982).

Modernization and urbanization are proceeding rapidly in many tropical countries. However, for workers to obtain good employment in modern cities, they must have a formal education and non-rural skills. It is extremely difficult for rural poor to be competitive for jobs in urban areas, because they lack the education and skills. Inevitably they become urban slum dwellers.

To solve these problems, widespread adoption of new development schemes is required, which significantly reduce under-employment and unemployment, and reduce rural–urban migration. Most rural workers would be happy to remain in the country, as long as there are employment opportunities in agriculture, forestry, or other rural occupations (Gajaseni, 1988). However, taungya reforestation as practised in Burma in the early 20th century is unsatisfactory for dealing with modern unemployment problems.

Taungya, as currently practised in Thailand, combines the advantages of reforestation and subsistence agriculture of the original Burmese colonial taungya system with modern experiments in social forestry in what is termed the Forest Village System. One of the social benefits of forestry projects as currently practised in Thailand and other Southeast Asian countries is their intensive use of labour. Table 3.1 gives figures which exemplify the requirement of labour for reforestation in Thailand. This employment increases income and community stability (Segustrom, 1976; Gregersen, 1982). Since the majority of the rural poor have few assets other than their capacity to work, they will benefit from reforestation projects.

There are many opportunities for workers to earn wages in the Forest Village Taungya System (Table 3.2). Weeding is an important activity in teak plantations for children up to 10–12 years of age. After that, at least three thinnings are required during a single rotation of teak. Finally, labour is needed during the final harvest of the teak.

Income from these activities is important in sustaining rural families and keeping them in the countryside. For example, in Indonesia, annual crops (grains) from modern taungya (tumpangsari) contribute only 20% of household revenue, but peasants can sustain their families from related plantation activities such as logging operations and collection of forest

**Table 3.1.** Labour requirements for reforestation in Thailand.

| Type of plantation | Labour required (man-days ha$^{-1}$) |
|---|---|
| Teak | |
| First year | 124.44 |
| Second year | 45.06 |
| Total | 169.50 |
| Non-teak | |
| First year | 142.25 |
| Second year | 58.06 |
| Total | 200.31 |

Source: Forest Industry Organization of Thailand.

**Table 3.2.** Opportunities for waged labour in teak reforestation in Thailand.

| Task | Wages paid (US$ $ha^{-1}$) |
|---|---|
| Establishment | 189.51 |
| Weeding, year 3 | 51.23 |
| Weeding, years 4, 5 (each) | 47.30 |
| Weeding, years 5–10 (each) | 18.63 |
| First thinning | 33.36 |
| Second thinning | 36.70 |
| Third thinning | 40.36 |

Source: Forest Industry Organization of Thailand.

products (Kartasubrata, 1986). Under-employment is less for those who participate in modern taungya.

## Productivity

It is possible to compare the amount of work required by a farmer to meet daily energy requirements under shifting cultivation and under taungya in Thailand. Grain has an energy value of about 4000 kcal $kg^{-1}$, and the energy requirement per person is about 2740 kcal per day (Han *et al.*, 1985). Under taungya, a farmer in 1 year will produce a quantity of grain that has a caloric value which can sustain 4.6 individuals for 1 year. In contrast, under shifting cultivation, a single farmer can produce the caloric requirements of 9.5 individuals for 1 year (Gajaseni, 1988). Clearly, under conditions in Thailand, shifting cultivation is more productive than taungya (Table 3.3). Therefore, unless there are definite socioeconomic advantages of taungya over shifting cultivation, conversion of shifting cultivation to taungya is unlikely to succeed in attracting farmers.

The socioeconomic advantage of taungya for peasants in Thailand is the wage from silvicultural activities. This is the factor that makes modern taungya more desirable.

Because of land scarcity in regions of Indonesia, taungya there is less productive on the basis of farm family than in Thailand. The average land allotted for taungya is only 0.25 ha per family, and this is sufficient only to sustain one individual for 1 year (Table 3.3). If the plot is fertilized, production improves to a level at which 2.5 individuals can be supported. Even with fertilization, there is not enough productivity to produce food for an average household of four individuals. Only if wages are available from plantation labour, such as logging and forest products' collection, can the average Indonesian family in taungya secure their annual energy requirement.

Use of fertilizers in taungya may be a dubious practice. While fertilizers do improve productivity, their costs are directly related to petroleum prices. For subsistence farmers to become dependent upon the petroleum-based economy could be harmful.

A report on taungya from Vietnam (Cuc, 1988) suggests that productivity there in a combination of rice and *Manglietia glauca* is highly productive (Table 3.3). However, the results must be interpreted cautiously. There are no available data on site quality, plot size, number of household members, and number of working hours per day.

## Forest Village Taungya

Lack of permanent settlement, under-employment, and low productivity are all problems for the rural poor which cannot be solved by the colonial approach to taungya. Some of the problems have been partially overcome by the modern Forest Village system of Thailand incorporating taungya. In this system the rural farmers live in villages subsidized by the government and private industry, and the families are given wages for silvicultural labour. They are permitted to grow crops during early years of the plantation.

In this section, social and economic aspects of this system are discussed.

**Table 3.3.** Comparison of grain production of agricultural systems in terms of number of individuals that can be supported from the production of one plot.

| Country | System | Area per family (ha) | Average no. individuals per family | No. individuals supported for 1 year |
|---|---|---|---|---|
| Thailand[1] | Shifting cultivation | 1.6 | 5.7 | 9.5 |
| Thailand | Taungya | 1.6 | 5.7 | 4.6 |
| Indonesia[2] | Taungya | 0.25 | 4.0 | 1.0 |
| Indonesia[3] | Taungya fertilizer | 0.25 | 4.0 | 2.5 |
| Vietnam[4] | Taungya | NA | NA | 5.4 |

[1]Modified from Gajaseni (1988)
[2]Modified from Kartasubrata (1986)
[3]Modified from Wiersum (1982)
[4]Modified from Cuc (1988)
NA No available data

## Rationale for peasants

Traditional agriculture, particularly shifting cultivation, has [illegible] as a cause of destruction of valuable forest timber. However, from the viewpoint of the rural farmer, the destruction of timber is irrelevant. The farmer will never receive any benefit from the timber. The majority of rural dwellers believe that the forest belongs to them, and they have the right to use it for whatever purpose they wish. To convince the peasant to stop cutting the forests for cultivation, the government must present a convincing economic argument. Based on experience in Burma and Sarawak, Leach (1957) pointed out that if there is an alternative which will improve income and standard of living, shifting cultivators will quickly adopt it. For the rural Thai peasant to join the Forest Village System, he must have an economic incentive, that is, he must be better off economically than he would be under shifting cultivation.

Income from grain production will not be sufficient to convince the farmer to join the Forest Village System. Based on income from grain production alone, a peasant will earn considerably more from traditional agriculture than from taungya in the Forest Village System (Table 3.4). The difference is substantial, and based on grain production alone, the forest village system is less appealing than traditional shifting cultivation. However, there are other factors which influence a farmer's decision to join the Forest Village System.

Many of the crops produced in shifting cultivation, and in the taungya of the forest village system are not sold, but consumed by the farmer and his family. Therefore, the price of grain produced is not as important to the peasant farmer as it would be to a commercial farmer. The Forest Village System provides wages for labour, thereby buffering the farmer against natural hazards such as floods or droughts.

Traditional farmers in Thailand are finding it increasingly difficult to find suitable land for agriculture. If land can be found, a land right transfer fee must be paid to the *de facto* owner, even if there is no legal private ownership (Gajaseni, 1988). By joining a Forest Village System, the farmer can earn money with which he hopes to buy a piece of land elsewhere.

Another important factor in convincing peasants to join the Forest Village System is government advertising and propaganda. Families may participate if they perceive the scheme to be in the interest of rural society and society as a whole.

## Rationale for enterprise economy

The main motivation for a farmer to participate in Forest Village taungya is reduction of risk, and stability of income. In contrast, for the entrepreneurs

**Table 3.4.** Costs and incomes for a peasant practising traditional agriculture (TA) compared to a member of the Forest Village System (FVS) at the end of the first year. Exchange rate was approximately 25 Baht per US dollar.

| Item | TA | | FVS | |
|---|---|---|---|---|
| | Total | Baht $ha^{-1}$ $year^{-1}$ | Total | Baht $ha^{-1}$ $year^{-1}$ |
| *Costs* | | | | |
| Housing | 1000 B (5 yr)[1] | 200.00 | None[4] | None |
| Land right transfer | 1250 B (60 yr)[1] | 20.83 | None | None |
| Tools | 228 B (10 yr)[1] | 22.80 | 228 B (10 yr)[1] | 22.80 |
| Planting stocks | 94.44 kg[2] | 566.67 | 77.78 kg[2] | 466.67 |
| Cattle manure | 111.11 kg | Free | 150.00 kg | Free |
| Total costs | | 810.30 | | 489.47 |
| *Incomes* | | | | |
| Grains[2] | 2366.67 kg | 14200.02 | 1160.00 kg | 6960.00 |
| Wage labour[3] | | None | 304.14 hr | 11405.30 |
| Total incomes | | 14200.02 | | 18365.30 |
| Net benefit | | 13389.72 | | 17875.83 |
| Income/cost | | 16.52 | | 36.52 |

[1]Longevity for depreciation
[2]6 Baht $kg^{-1}$
[3]37.5 Baht 8 $h^{-1}$
[4]FIO will provide all materials needed

that finance the system, the motivation is monetary gain. Even where the government backs the Forest Village System, non-monetary revenues such as social development are secondary to balancing international trade through production of valuable timber. Forest Village taungya has never been justified solely by the improvement of the socioeconomic status of the rural poor.

For an enterprise economy establishing forest plantations, taungya gives satisfactory results. It is one of the best management strategies to reduce the cost of reforestation. Costs for teak reforestation by taungya are given in Table 3.5.

Table 3.6 is an analysis of the benefit/cost ratio for reforestation by taungya. The ratio for taungya is favourable for long rotations in Thailand, and for short rotations in Nigeria. The ratio is higher than for reforestation using permanent or casual labour (Table 3.6). Even when the benefits are discounted at 9% over the life of the rotation, the net present value in taungya systems is higher than the cost (Table 3.6). This indicates an economically efficient system (Gregersen and Contreras, 1979). Thus, the taungya system for timber production appears to be a very effective means for private entrepreneurs or enterprises to make profits.

## Improvements Needed

### Socioeconomic

Integrated forest management, which combines national objectives and community requirements, is needed for further development and improvement of tropical rural areas (Eighth World Forestry Congress, 1978). The

**Table 3.5.** Cost for teak reforestation.

| Item | Cost (US$ $ha^{-1}$) | Percentage of total |
|---|---|---|
| Establishment labour | 189.51 | 29 |
| Thinning | 110.42 | 17 |
| Infrastructure[1] | 47.10 | 7 |
| Non-periodic[2] | 306.13 | 47 |
| Total | 653.16 | 100 |

Source: Boonkird *et al.*, 1984.
[1]Properties, maintenance, management.
[2]Primarily weeding.

**Table 3.6.** Economic aspects of reforestation enterprise. Figures are US$ $ha^{-1}$.

| | Cost | Benefit | Benefit/cost | Net present value (9% discount) |
|---|---|---|---|---|
| *Thailand*[1] (teak + upland rice, 60-year rotation) | | | | |
| 1. Labour included | 568.46 | 2172.91 | 3.82 | 1604.45 |
| 2. Labour excluded (taungya) | 378.95 | 2172.91 | 5.73 | 1793.93 |
| *Nigeria*[2] (*Gmelina* + food crops, 8-year rotation) | | | | |
| 1. Permanent labour | 400.00 | 359.84 | 0.90 | −40.16 |
| 2. Casual labour | 305.50 | 359.84 | 1.18 | 54.34 |
| 3. Taungya | 213.85 | 359.84 | 1.68 | 145.99 |

[1]From Gajaseni (1988).
[2]From Nwonwu (1987). For Nigeria, permanent labour was calculated at US$1.56 per man-day, and casual labour at US$1.25 per man-day.

World Bank has announced a major shift in its support of forestry activities, and will now place more emphasis on the environment and rural forestry (Spears, 1982).

Few improvements have occurred, however. Even though taungya has provided some socioeconomic advantages to peasants over shifting cultivation, and to business enterprise for reforestation, the system could still be considerably improved to contribute more toward rural development. Improvements should include aspects of permanent settlement, increasing income, and improving living conditions.

Permanent settlement should be considered as an integral part of improvement schemes. The strategies must be focused upon enhancing the intimate relationship between people and the land. However, the problem of land tenure and land ownership is one that must be approached cautiously.

Further opportunities for increasing income and increasing the stability of income must also be developed. Lack of opportunity for waged labour has resulted in a decrease in the number of families in the Forest Village at Mae Moh, Thailand, studied by Gajaseni (1988). The number of villagers reached the highest number in 1982, with 105 families and 528 individuals. However in 1985, 20 families emigrated back to their original residences in the northeastern provinces. The reason, as told to the village council, was shortage of waged labour from the Forest Industry Organization. Because of the lack of land to be cultivated for new plantations, the officer-in-charge had to reduce working days for all families. All members suffered from this. This experience re-emphasizes that monetary support is the most important key to success of the Forest Village Enterprise.

Living conditions of farmers in taungya projects should be improved, and made comparable with life under urban conditions. Otherwise, the peasants will abandon the system and migrate to the cities. Basic facilities such as schools and public health centres are critical (EGAT, 1982).

Implementing these socioeconomic improvements will necessarily decrease to a certain extent the profitability of the timber-production enterprise. However, such a sacrifice may be necessary to ensure the survival, continuity, and long-term profitability of the system.

### Cultural

Rural development programmes frequently encounter, and are often specifically aimed at, groups of rural people who possess cultural beliefs and values different from those of the people who originally designed the development programme or who are carrying it out. Many government-organized agricultural development schemes on Peninsular Malaysia are not encouraging, because of the incompatibility between the traditional nomadic cultural values of the people to be settled, and the demands of settled life within the system (Endicott, 1979).

## Conclusion

In tropical countries, people living in or near the forest are usually part of the poorest and least powerful sectors. They are socially and economically marginal and face increasing socioeconomic stress due to lack of legal or political support.

Economic development in rural tropical areas has been largely extractive and damaging, as exemplified by lumbering with little or no reforestation. Some forestry projects have been developed without consideration of the local inhabitants, or sometimes even in opposition to their wishes. Traditional systems of agriculture and subsistence may work well or badly. Nevertheless, it is irresponsible to ignore or suppress outcries against elimination of traditional agriculture without providing substitutes of proven practicality (Gajaseni and Jordan, 1990).

The socioeconomic aspects of development in the rural tropics encourage the support of Forest Village taungya. The system provides a realistic opportunity in which rural people can participate. It is compatible with prevailing local values and customs. However, improvements still are required if the system is to be permanently successful. The three major advances needed are:

1. ensuring permanent settlement;
2. increasing income earning opportunities;
3. developing supportive facilities such as schools and health centres.

In Thailand, many of these improvements can be carried out within the existing framework of the Forest Village System of Taungya.

## References


Ariki, S. (1986) Comparative studies on agroforestry and forestedland ownership in Japan, Thailand, and Indonesia. *Comparative Studies on the Utilization and Conservation of the Natural Environment by Agroforestry Systems*, pp. 327–341.

Ashton, P.S. (1977) The UNESCO state of knowledge report on tropical forest ecosystems: summary review and suggested emphasis for future research. In Bruning, E.E. (ed.) *International MAB-IUFRO Workshop on Tropical Rain Forest Ecosystem Research*, pp. 33–37.

Banijbatana, D. (1978) Forestry policy in northern Thailand. In Kunstadter, P., Chapman, E.C. and Sabhasri, S. (eds) *Farmers in the Forest.* The University of Hawaii Press, Honolulu. pp. 54–60.

Boonkird, S., Fernandes, E.C.M. and Nair, P.K.R. (1984) Forest villages: An agroforestry approach to rehabilitating forest land degraded by shifting cultivation in Thailand. *Agroforestry Systems* 2, 87–102.

Budowski, G. (1982) The socio-economic effects of forest management on the lives of people living in the area: the case of Central America and some Carribbean countries. In Hallworth, E.G. (ed.) *Socio-economic Effects and Constraints in Tropical Forest Management.* John Wiley and Sons, New York, pp. 87–101.

Cuc, L.T. (1988) *Agroforestry Practices in Vietnam.* Working Paper No. 9, East–West Center, Honolulu, Hawaii.

EGAT (1982) *Mae Moh Project: Environment and Ecological Studies and Impact Assessment.* Electrical Generating Authority of Thailand.

Eighth World Forestry Congress (1978) *Proceedings.* Jakarta, Indonesia. International Union of Forest Resource Organizations, and FAO, Rome.

Enabor, E.E., Okojie, J.A. and Verinumbe, I. (1982) Taungya system: Socio-economic prospects and limitations. In Macdonald, L.H. (ed.) *Proceedings of the Workshop on Agro-forestry in the Humid Tropics.* The United Nations University, Tokyo, Japan, pp. 59–64.

Endicott, K. (1979) The impact of economic modernization of the Orang Asli (aborigines) of northern Peninsular Malaysia. In Jackson, J.J. and Rudner, M. (eds) *Issues in Malaysian Development.* Heinemann, Singapore, pp. 167–204.

FAO (1984) *Land Evaluation for Forestry.* FAO Forestry paper 48. FAO, Rome.

Gajaseni, J. (1988) *Ecological Comparison of Traditional Agriculture and the Forest Village System (Agroforestry) in Northern Thailand.* PhD Dissertation, University of Georgia, USA.

Gajaseni, J. and Jordan, C.F. (1990) *Diversified Agroforestry System.* BIOTROP Special Publication 39: Agroforestry Systems and Technologies, pp. 157–161.

Gregersen, H.M. (1982) Environmental constrains versus economic gain in tropical forestry. In Mergen, F. (ed.) *The International Symposium on Tropical Forests Utilization and Conservation.* Yale University, Connecticut, pp. 108–124.

Gregersen, H.M. and Contreras, A.H. (1979) *Economic Analysis of Forestry Projects.* FAO Forestry Paper 17. FAO, Rome.

Han, C.R., Golley, F.B. and Mou, Z.G. (1985) Energy analysis of advanced collective farms in north China. *Agriculture, Ecosystems & Environment* 13, 217–240.

Kartasubrata, J. (1986) Agroforestry in Indonesia with special reference to Tumpangsari in forest area. *Comparative Studies on the Utilization and Conservation of the Natural Environment by Agroforestry Systems,* pp. 29–51.

Kunstadter, P. and Chapman, E.C. (1978) Problems of shifting cultivation and economic development in northern Thailand. In Kunstadter, P., Chapman, E.C. and Sabhasri, S. (eds) *Farmers in the Forest.* The University of Hawaii Press, Honolulu, pp. 3–23.

Leach, E.R. (1957) Some economic advantages of shifting cultivation. *Proceedings of the Ninth Pacific Science Congress,* Vol. 7. Chulalongkorn University, Thailand.

Nwonwu, F.O.C. (1987) Cost minimization through the use of taungya system in pulpwood plantation establishment. *Agroforestry Systems* 5, 455–462.

Segustrom, G. (1976) Creating work and caring about workers in tropical countries. *Unasylva* 114, 2–9.

Seibert, B. (1990) *Aspects of Agroforestry for Lowland Humid Areas with Low Population Densities, The Example of East Kalimantan, Indonesia.* BIOTROP Special Publication 39: Agroforestry System and Technologies, pp. 147–155.

Spears, J. (1982) Sustainable land use systems. In Mergen, F. (ed.) *Proceedings of the International Symposium on Tropical Forest Utilization and Conservation.* Yale University, Connecticut, pp. 14–47.

Vergara, N.T., Tingsabadhi, C., Johnson, K., Vidyarthi, V. and Bowen. R. (1986) *Social Forestry Research Issues: Preliminary Problem Identification in Sisaket Province, Northeast Thailand.* ODI Network Paper 2b.

Wiersum, K.F. (1982) Tree gardening and taungya on Java: Examples of agroforestry techniques in the humid tropics. *Agroforestry Systems* 1, 53–70.

World Bank (1982) *World Bank Operations.* The Johns Hopkins University Press, Baltimore.

Chapter 4

# Tree–Crop Interactions in Taungya Plantations

Hiroyuki Watanabe

*Laboratory of Forest Resources, Faculty of Agriculture, Kyoto University, Kyoto 606, Japan*

For the last two decades, the taungya system of forest plantations has been initiated in many regions of Southeast Asia to rehabilitate land areas degraded by activities such as shifting cultivation or intensive logging. Taungya has been applied in many regions where it was never traditionally practised. Many combinations of trees and crops are found in the various taungya systems of Asia. The combinations are dictated by site conditions such as climate, soil fertility, and topography, as well as socioeconomic conditions such as market demand and skills and preferences of cultivators (Watanabe *et al.*, 1988). Often, scant attention is paid to potential interactions when combinations of trees and crops are selected. As a result, a number of agronomic and silvicultural problems have arisen that result from interactions between crop plants and trees. The purpose of this chapter is to discuss these interactions, and evaluate them in terms of how they affect the taungya system.

## Relationship between Trees and Crops

Filius (1982) recognized three types of interactions between trees and crops in agroforestry systems: complementary; supplementary; competitive.

Complementarity exists if the presence of one crop increases production of another crop. An example of complementarity in cropping is the positive effect of one crop on the other crop. *Cordia alliodora* and several *Erythrina* species are used as shade trees for coffee and cocoa. Trees moderate the intensity of sunlight and wind, and maintain higher humidity. Crops under or between the trees are protected from sudden changes in climate. Species of *Acacia, Alnus, Leucena, Gliricidia* and other legumes are often interplanted with crops in agroforestry systems, because the nitrogen fixed by these plants increases soil fertility and benefits the crop plants.

Interplanting trees and crops as in the taungya system often results in a mutual benefit. Weeding carried out by cultivators benefits both the crops and the trees in that it reduces competition for nutrients and water. The cut weeds represent a source of readily available nutrients to the crops and trees, if used as a mulch. The crops and weeds, and their litter when they are cut provide a mat over the soil which lessens erosion and reduces evaporation.

If the presence of one crop does not influence production of other crop(s), we have supplementarity, an independent relationship (Fillius, 1982). This relationship occurs if the different crops draw on resources at different times of the year, or from different parts of the environment, e.g. different soil depths for nutrients. In taungya, crop plants occupy space between tree seedlings, and use light, nutrients and water that presumably otherwise would be wasted. During the first growing season, tree seedlings are so small that crops usually grow well despite their presence. The relationship between trees and crops during the first year of taungya appears in most cases to be supplementary.

## Negative Interactions

### Competition

Ideally, the relationship between crops and trees should be of a 'complementary' nature. However, this is not always the case. In many situations where trees and crops are grown together, they may compete for water, nutrients and solar energy. The situation is obvious when the canopies of trees begin to close over the tops of crops such as upland rice.

In Indonesian taungya (Tumpangsari), teak is often interplanted with giant ipil-ipil (*Leucaena leucocephala*) and with cassava as a cash crop (Fig. 4.1). The ipil-ipil is beneficial in that it produces shade and green manure. The side-shading results in good form for the teak. However, competition can be deleterious when teak is enclosed by two rows of ipil-ipil (Coster *et al.*, cited in Kartasubrata, 1986).

In some regions such as Java, perennials are being interplanted in systems of taungya (see Chapter 10). Competition between trees and crops is a long-term problem in plantations when the crop species is a perennial. Such combinations also occur in peninsular Thailand, where perennial crops such as coffee, cashewnut, or other fruit trees are commonly planted

**Fig. 4.1.** In the Indonesian taungya (Tumpangsari), teak is often interplanted with the giant ipil-ipil which provides shade and green manure. Cassava is usually the agricultural crop. The figure suggests that competition is occurring.

together with para-rubber (*Hevea brasiliensis*) or Polynesian wood (*Casuarina equisetifolia*) (Fig. 4.2).

## Allelopathy

Allelopathy refers to plant–plant biochemical interactions that have detrimental effects. Allelopathic compounds are released into the environment by means of volatilization, leaching, decomposition of residues and root exudation (Chou, 1990).

Allelopathic interactions between trees and crops have recently been

**Fig. 4.2.** Competition is probably occurring in this system where the perennial crop cashewnut (*Anacardium occidentale*) is interplanted between polynesian wood (*Casuarina equisetifolia*) in peninsular Thailand.

investigated in taungya plantations. Susesh and Vinaya Rai (1987) reported on effects upon sorghum, cowpea, and sunflower by blue gum (*Eucalyptus tereticornis*), Polynesian wood (*Casuarina equisetifolia*) and giant ipil-ipil (*Leucaena leucocephala*). The crops, grown in top soil taken from beneath the trees, had decreased seed germination rate, root growth, and dry matter production compared with crops grown in control soils (Table 4.1).

Del Moral and Muller (1970) have discussed evidence that trees such as *Eucalyptus* spp. and *Leucaena leucocephala* produce toxins that inhibit the growth and decrease yields of crops interplanted with these species.

**Table 4.1.** Response of test crops to different tree crops.

| Tree species | Germination (%) | | | Root length (cm) | | | Dry matter production (mg plant$^{-1}$) | | |
|---|---|---|---|---|---|---|---|---|---|
| | So | Co | Su | So | Co | Su | So | Co | Su |
| *Eucalyptus* | 54.1 | 85.0 | 86.6 | 7.1 | 7.5 | 7.0 | 377 | 1188 | 963 |
| *Casuarina* | 73.3 | 84.1 | 82.5 | 8.0 | 8.2 | 7.5 | 519 | 1506 | 1242 |
| *Leucaena* | 58.3 | 89.1 | 89.1 | 8.4 | 8.0 | 7.8 | 611 | 1506 | 1151 |
| Control | 80.0 | 96.6 | 93.3 | 7.2 | 9.3 | 8.7 | 1328 | 2015 | 1422 |
| | | sed | CD | | sed | CD | | sed | CD |
| Crop | | 2.2 | 4.3 | | 0.08 | 0.16 | | 39 | 77 |
| T × C | | 6.2 | 12.2 | | 0.16 | 0.33 | | 78 | 154 |

Source: Suresh and Vinaya Rai, 1987.
So, sorghum; Co, cowpea; Su, sunflower.
sed: standard error of deviation; CD: coefficient of determination.

Chou and Yang (1982) and Chou and Kuo (1986) have identified the allelopathic compounds produced by *Leucaena leucocephala*, *Phyllostachys edulis*, and *Cryptomeria japonica* that inhibit growth of other species.

It is usually difficult under field conditions to distinguish between allelopathic effects and competition for light, nutrients, and water. In northeast Thailand, crops such as cassava, kenaf, and groundnut have lower yield when grown in combination with *Eucalyptus camaldulensis* than when grown alone (Fig. 4.3). The smaller groundnut plants beneath the eucalyptus could be due to competition or to allelopathic effects.

## Conflict between Landowners and Cultivators

Competition and negative interactions that occur between trees and crops give rise to conflicts between landowners who tend trees and cultivators who grow crops.

Kartasubrata (1986) has reported that Tumpangsari results in a decline in the growth of young teak trees. However, the decline may be due to actions of the peasant, rather than to direct interactions. Decline in tree growth occurs when the farmer cuts back the trees to lessen competition with his crops. For example, pineapple as a cash crop is commonly cultivated with teak (*Tectona grandis*) in northern Thailand, and with falcate-leaved albizia (*Paraserianthes falcata* = *Albizia falcataria*) in central Java.

**Fig. 4.3.** Groundnuts interplanted in a stand of *Eucalyptus camaldulensis*. Allelopathic effect on groundnuts is suspected.

To allow sufficient sunlight to reach the pineapple, cultivators drastically cut back the branches of the trees (overlopping) (Fig. 4.4). Sometimes even the tops of the trees are cut. This results in undesirable reduced growth of the trees.

Competition for light is also recognized in teak plantations with upland rice in east and central Java, and in red gum (*Eucalyptus camaldulensis*) or Burmese rosewood (*Pterocarpus macrocarpus*) plantations with cassava in northeast Thailand. The overlopping of the trees results in a sparse stand of trees with strange form (Fig. 4.5).

**Fig. 4.4.** Pineapple interplanted amongst teak in taungya in northern Thailand. The branches of the teak are severely lopped, to ensure adequate sunlight reaches the pineapple.

Severe pruning or overlopping of trees to increase light available to crops is common when the crops are vegetables which require much sunlight. In high altitude plantations in Java, Indonesia, cabbage, onions, potatoes and other vegetables are cultivated between pine (*Pinus merkusii*) which is severely pruned back (Fig. 4.6).

Cultivators usually prefer to cultivate a plot as long as possible, because of the high labour costs in preparing and weeding newly reclaimed land. They would prefer to remain on the same land as long as there is no large reduction in harvest. However, landowners interested primarily in tree production prefer to have cultivation cease after one or two years.

**Fig. 4.5.** Burmese rosewood (*Pterocarpus macrocarpus*) showing form influenced by severe overlopping.

From the point of view of the forester whose main interest is in maximizing wood production, the tendency to prune back trees to favour crops is undesirable. Overlopping or pollarding results in abnormal form of trees, from which there is little hope of recovery. To favour growth of trees, crown closure of trees should be accelerated because this reduces soil erosion, reduces competition from weeds, and minimizes the invasion of fire in the dry season, a factor especially important in the monsoon region.

In highly populated regions of Indonesia such as Java, where agricultural land is limited, this conflict can be a major problem. There are many people eager to join taungya reforestation projects. To minimize conflicts

**Fig. 4.6.** Vegetable crops cultivated in a pine (*Pinus merkusii*) plantation at high altitude in Java.

between farmers and tree-growers, cultivation of perennial crops and vines is often prohibited, and cultivation of annuals is allowed for only 2 years.

In Thailand, cultivators are required to shift their site of cultivation after a maximum of 3 years. This is due to competition between trees and crops, but it is also due to the need by the government for labour to prepare new land for plantations, and to plant seedlings. This sometimes makes taungya unattractive for peasants, who prefer to remain in the same location and cultivate the same area for longer periods. Because of this reluctance on the part of peasants, the Forest Village System of Thailand has not always accomplished the reforestation that was planned (Phothitai, 1984).

An approach that can be used to induce the peasant farmer to give better care to the trees is to give the farmer a bonus for every tree that reached a certain height or coverage within a specified period of time. In the Philippines, the farmers actually own the trees, and profit by selling them to local pulp mills (see Chapter 11).

## Spacing

Spacing of trees is at the heart of the matter in the dispute between cultivators and foresters. Cultivators prefer the widest, least dense spacing of trees possible, to give advantage to the crops. Spacing for teak plantations by the taungya method was originally 4 × 4 m (625 trees $ha^{-1}$). Recently there have been experiments with spacing of 2 × 8 m. This results in the same density of trees, while allowing easier cultivation of crops.

Spacing in Thailand is wider than the 1.8 × 1.8 m intervals used when taungya was established in Burma (Blanford, 1925). Wider spacing results in more infestation by tall grasses such as alang-alang (*Imperata cylindrica*), *Pennisetum* spp., and various species of bamboo. These grasses compete with the trees, and also make the plantations susceptible to fire which can severely damage or kill the trees. Weeding for many years is necessary to prevent competition and fires. Weeding as a fire control measure in Thailand is illustrated in Fig. 4.7.

To ensure successful forest regeneration and recovery of wasteland, closer spacing and denser spacing is recommended. However, the importance of cultivators and their need for wider spacing for crop cultivation in taungya is also recognized. There is a need for further research on spacing and tree–crop combinations, to optimize benefits for both the farmer and the forester.

Interactions between crops and trees in agroforestry systems has already received attention (e.g. Huxley, 1983). However, optimal spacings and combinations of crops and trees are frequently site-specific, and depend on local evironmental conditions such as soil, slope, and weather, as well as local economic conditions (market demand), cultural conditions (preferred agricultural tradition), and political situations (land ownership regulations).

## Conclusion

The taungya method has promise of being an effective reforestation measure to rehabilitate degraded, over-used land, and to give employment to landless people. However, the taungya system is far from ideal, because of the competition that occurs. Competition can be intense between trees

**Fig. 4.7.** Even in mature stands, weeding may be necessary to prevent fire. This photograph illustrates dry weeds in a stand of *Melia azedarach*.

and crops, and the people who want to grow trees and the people who want to grow crops. There is ample room for research to determine combinations of crops and trees where competition is minimized. This research should focus both on the combinations of species and their physiological requirements, and also on the spacing of the trees.

**1.** Studies to determine optimum combinations of trees and crops. Shade tolerant crops could be more desirable than species which always need full sunlight. Trees with thin canopies would be more compatible with crops than trees with thick canopies.
**2.** Studies to determine optimum spacing of trees. Optima must be defined

for each situation, and involves a balance between wood production, crop production, and costs of thinning, weeding, and pruning.

**3.** Studies to determine best methods of pruning and thinning. Some pruning and thinning is necessary to obtain crop yield and good shape and growth of the trees themselves. However, overpruning, overlopping, and overthinning can be deleterious.

## References


Blanford, H.R. (1925) Regeneration with the assistance of taungya in Burma. *Indian Forest Records* xi, Part III, 41 pp.

Chou, C. (1990) The role of allelopathy in agroecosystems: studies from tropical Taiwan. In Gliessman, S.R. (ed.) *Agroecology: Researching the Ecological Basis for Sustainable Agriculture.* Ecological Studies 78. Springer Verlag, New York, pp. 104–121.

Chou, C. and Kuo, Y. (1986) Allelopathic research of subtropical vegetation in Taiwan III. Allelopathic exclusion of understory by *Leucaena leucocephala* (Lam.) de Wit. *Journal of Chemical Ecology* 12, 1431–1448.

Chou, C. and Yang, C. (1982) Allelopathic research of subtropical vegetation in Taiwan II. Comparative exclusion of understory by *Phyllostachys edulis* and *Cryptomeria japonica. Journal of Chemical Ecology* 8, 1489–1507.

Coster, C., Hardjowasono and Soetarmo, M. Agricultural crops in teak plantations. *Tectona* 28, 464–487. Cited in Kartasubrata (1986).

Huxley, P.A. (ed.) (1983) *Plant Research and Agroforestry.* International Council for Research in Agroforestry. Nairobi, Kenya.

del Moral, R. and Muller, C.H. (1970) The allelopathic effects of *Eucalyptus camaldulensis. American Midland Naturalist* 83, 254–282.

Filius, A.M. (1982) Economic aspects of agroforestry. *Agroforestry Systems* 1, 29–39.

Kartasubrata, J. (1986) *Agroforestry in Indonesia with Special Reference to Tumpangsari in forest areas.* Monsoon Asia Agroforestry Joint Research Team: Comparative studies on utilization and conservation of the natural environment by agroforestry systems, pp. 29–51.

Phothitai, M. (1984) *Agroforestry practiced by the Forest Industry Organization (FIO).* Bangkok, 20 pp.

Suresh, K.K. and Vinaya Rai, R.S. (1987) Studies on the allelopathic effects of some agroforestry tree crops. *International Tree Crops Journal* 4, 109–115.

Watanabe, H., Sahunalu, P. and Khemnark, C. (1988) Combination of trees and crops in the taungya method as applied in Thailand. *Agroforestry Systems* 6, 169–177.

Chapter 5

# Nutrient Cycling in Taungya

Carl F. Jordan

*Institute of Ecology, University of Georgia, Athens, Georgia 30602, USA*

Soils of the lowland tropics and subtropics are usually more highly weathered and nutrient-deficient than soils of higher latitudes (Sanchez, 1976). Natural tropical ecosystems have evolved mechanisms to conserve nutrients, such as efficient transfer of nutrients from decomposing litter into roots at or near the soil surface. These mechanisms can be destroyed by management such as logging, farming and ranching (Jordan, 1982). As a result, the effect of management on nutrient dynamics in tropical ecological systems is frequently more severe than in higher latitude systems (Jordan, 1985).

One objective of this book is to evaluate taungya in terms of its sustainability. Since an adequate supply of nutrients is essential to a sustainable system of taungya, it is important to evaluate the effect of management during taungya on nutrient stocks and losses.

There are two phases during a cycle of taungya plantation when the system is particularly susceptible to nutrient loss. The first is immediately after the forest is cut and burned and during the first years of cultivation. That phase is considered in this chapter. The second is during the harvest of the trees at the end of the rotation (Aboriside and Aweto, 1990). This problem is addressed by Bruijnzeel in the next chapter.

There have been few studies of changes in soil properties during the initial cultivation stage. Tulaphitak *et al.* (1983) studied the effect of shifting cultivation on the physical and chemical properties of soil at a site in eastern Thailand with a monsoonal climate. Although the study site itself was not part of a taungya system, it was in the region where taungya was practised, and cultivation of maize within their system was similar to cultivation of maize within taungya of Thailand. They stated (Tulaphitak *et al.*, 1983):

> Heavy initial loss of ash and soil are caused by a few premonsoon rains. In shifting cultivation in the monsoon tropics, burning of the felled trees must be done towards the end of the dry season and that is a difficult time to establish a crop cover. Only with more assured rains can crops be sown or planted. Therefore, there is a period very susceptible to soil erosion between the time of burning and the time of crop establishment. It appears inevitable for shifting cultivation in the monsoon tropics to suffer from the severe initial loss of nutrients.

At the same site that Tulaphitak *et al.* (1983) carried out their study, Takahashi *et al.* (1983) examined the effect of different cultivation techniques on soil erosion and found that the greater soil cover resulting from intercropping of mixed crops with different seeding times, or of perennial crops and annual crops significantly reduced erosion. Alexander (1987) working in Kerala, India found that

> The site shock, including accelerated erosion generated during the establishment phase by land clearing and preplanting operations can be counteracted through protective and ameliorative measures provided by ground cover. Taungya provides an early soil cover with trees and annual crops badly needed under the highly erosive rains, thus minimizing nutrient loss. However, its effectiveness depends on the nature of the crops and cultural practices.

We do not know of any studies on nutrient dynamics during the initial stage of taungya. There have been, however, studies of nutrient cycling in systems of shifting cultivation. In shifting cultivation as in taungya, the period during which grain or root crops are planted is 1–3 years. In the following section we examine the effect of shifting cultivation on depletion of nutrient stocks, and its significance for subsequent growth of trees. These studies may shed light on the question of whether there are nutrient losses during initial stages of taungya that can effect subsequent growth of trees.

## Nutrient Dynamics During Shifting Cultivation

Nutrient loss occurred during shifting cultivation in Ghana (Nye and Greenland, 1964), Thailand (Zinke *et al.*, 1978), Peru (Denevan, 1971; Sanchez *et al.*, 1983), Venezuela (Harris, 1971) and Brazil (Brinkmann and Nascimento, 1973).

Some of the loss is due to rapid leaching of nutrients originally in the biomass of the forest and subsequently converted to ash. This was found to occur on an Oxisol soil in the Amazon Territory of Venezuela near San Carlos (Jordan, 1989). That study compared nutrient leaching in two plots that were cut and burned, with leaching from a control plot. Of the two plots that were cut and burned, one was cultivated in the traditional method and planted with manioc and plantain, and the other was allowed to regenerate to secondary successional vegetation.

Most nutrients in the San Carlos study showed some response to cutting and burning, but the most dramatic increases were in potassium, magnesium and nitrate nitrogen (Fig. 5.1). There were no consistent differences in leaching between the shifting cultivation plot and the plot abandoned to succession.

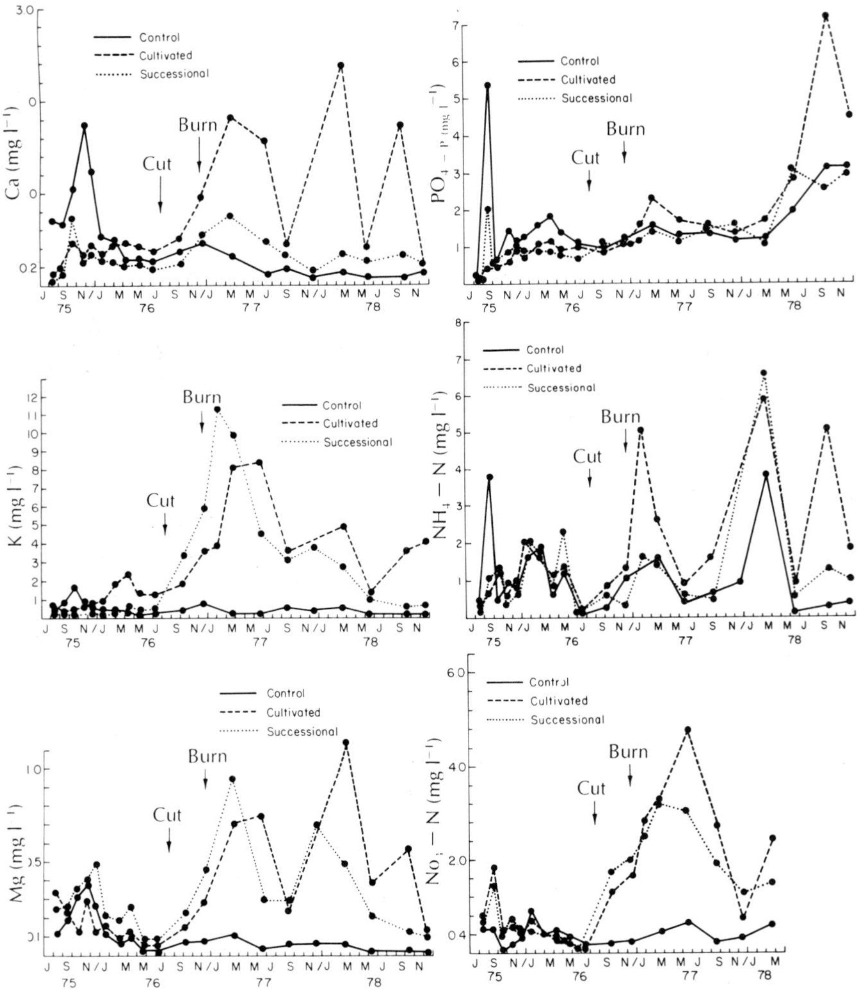


**Fig. 5.1.** Concentrations of nutrients in leachate from 12 cm depth of the experimental and control plots of the shifting cultivation experiment at San Carlos de Rio Negro, Venezuela.

Although the increase in leaching rate of some nutrients was striking, these data alone allow no conclusions on the implications of the losses for future uptake by trees after disturbance has ceased. Total stocks remaining in the ecosystem must be quantified to evaluate the relative impact of the losses.

Total stocks in the undisturbed forest at San Carlos, and the changes caused by cutting, burning and cultivation are shown in Fig. 5.2. More than half of the calcium, potassium and magnesium in the ecosystem were held in the woody biomass of the forest. When the trees were cut and burned, most of these cations were converted into ash. During the 3 years of cultivation, about half of the potassium and about one-third of the calcium and magnesium in the system were lost through harvest or leaching. Nevertheless, at the time cultivation was abandoned, total stocks of these nutrients

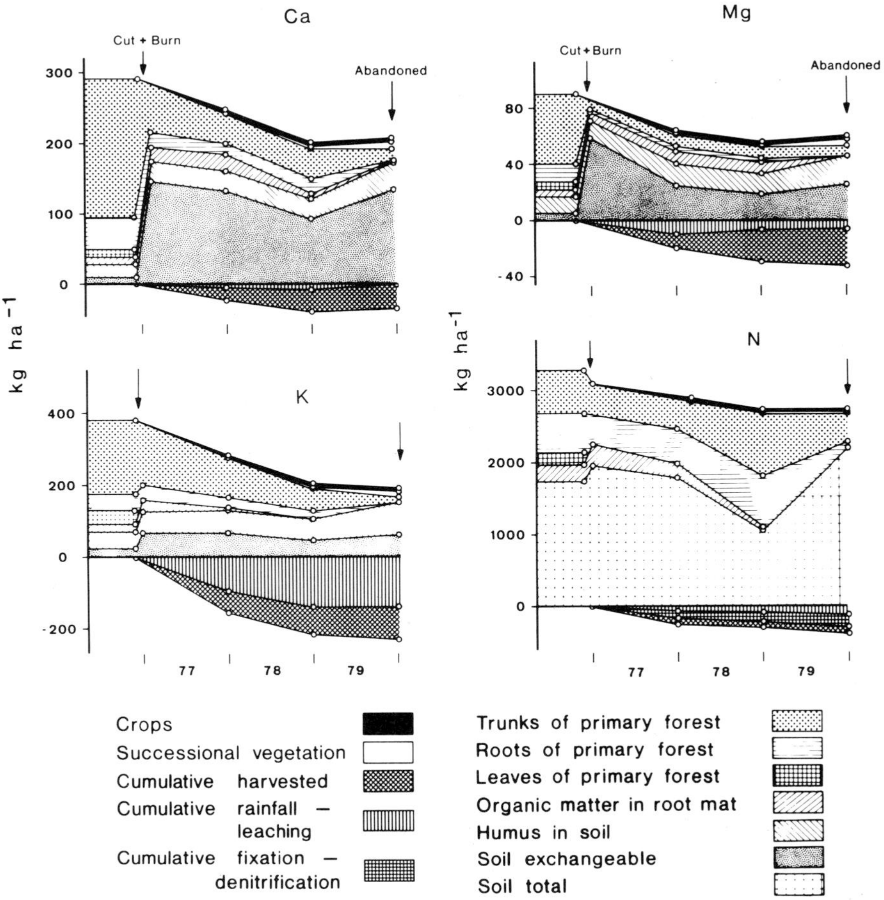


**Fig. 5.2.** Stocks and cumulative losses of calcium, potassium, magnesium and nitrogen in the shifting cultivation plot at San Carlos de Rio Negro, Venezuela.

in the soil were still many times higher than the stocks of the original forest.

In contrast to calcium, potassium and magnesium, more than three-quarters of the nitrogen in the ecosystem was in the soil, associated with living and non-living below-ground organic matter, including roots, and humus. As a result of the burn and subsequent cultivation, less than 10% of the nitrogen was lost. Eighty four percent of the phosphorus in the ecosystem was in the soil. The soils at the study site were very high in iron and aluminium, as are many of the soils of the lowland tropics. As a result, almost all of the phosphorus in the soil was bound with the iron and aluminium in an insoluble state. There was no detectable loss of total phosphorus from the ecosystem as a result of forest clearing, burning, or cultivation.

Despite the relatively high amount of total nutrients in the soil during the period of cultivation, total dry weight net primary productivity dropped from 5.6 t $ha^{-1}$ per year the first year to 4.1 t $ha^{-1}$ per year the third year. Edible crop production dropped from 3.5 t $ha^{-1}$ per year the first year to 1.8 t $ha^{-1}$ per year the third year (1989). However, the first year following abandonment of cultivation, the net production of the secondary vegetation was 7.2 t $ha^{-1}$ per year, a value higher than that of the undisturbed forest, and by the third year, the rate was 14.1 t $ha^{-1}$ per year (Jordan, 1989). It appears that the nutrient loss due to cultivation at the San Carlos site had little, if any, effect on the subsequent productivity of the trees.

If there were sufficient nutrients in the soil to permit vigorous growth of secondary trees, what was the cause of the decline in crop production? Weeds, insects, and disease were ruled out as contributing factors. A series of studies on the status of phosphorus in the soil led to the hypothesis that it was not the amount of phosphorus in the soil that affected crop growth, but rather the state of the phosphorus. During the cultivation of crops, most of the available phosphorus in the soil became fixed by the iron and aluminium.

What controlled the availability of phosphorus? During the process of decomposition of organic matter, either in the humus on top of the soil or the organic matter in the soil, organic acids such as malic and citric acid are released. They may be released from the material being decomposed, or by metabolism of the microbes carrying out the decomposition. These organic acids can reduce fixation of phosphate by iron and aluminium (Lee *et al.*, 1990). Some acids such as piscidic acid (*p*-methoxybenzyl tartaric acid) may have the capability of solubilizing phosphate already bound in insoluble compounds (Ae *et al.*, 1990). In the undisturbed forest, these organic acids may play an important role in recycling phosphorus. The gradual disappearance of humus on the soil surface during the period of cultivation may have resulted in a lessening of the amount of phosphorus in an available state, and as a consequence, crop productivity decreased. Phosphorus fixation seemed to be the most immediate cause of problems in soil fertility during cultivation of crops.

The good growth rate of successional trees in the plot compared to the low production of the crop species may be related to the better ability of wild species to acquire recalcitrant phosphorus. This could be achieved through mycorrhizae, or it could be related to the build-up of a litter layer and humus on the soil. As this material decomposes, organic acids that are released may be important in making phosphorus more soluble.

The soil at the study site was classified as a petroferric gibbsiothox (Dubroeucq and Sanchez, 1981), a type of oxisol extremely low in nutrients and nutrient-holding capacity. Except for sites on podsol soils, the site probably represents one of the most nutrient-poor situations in the lowland tropics. From the aspect of nutrient cycling, it represents a system in which effects of nutrient loss would be most likely to occur. The apparent lack of effect of 3 years shifting agriculture would suggest that nutrient loss during taungya would ordinarily not be a serious problem.

Do the trees affect the nutrition of the crop during taungya? Verinumbe and Okali (1985), studying the effect of first-year teak sprouts on nutrients in soil and interplanted maize, concluded: 'It is suggested that soil nutrient changes due to canopy or surface roots of coppiced teak regrowth are probably too small to influence the performance on intercropped maize in the first season following coppicing.' Eventually, as the trees increase in size, they compete with crops for resources. It is for this reason that farmers sometimes cut back the trees, a problem mentioned by Watanabe in Chapter 4 of this book.

## Conclusion

Cutting and burning the forest results in conversion of some of the nutrient stocks of forest ecosystems from a bound state in the biomass into soluble ash. Cultivation results in a loss of some of these stocks, through erosion, leaching, volatilization or fixation. Although there are losses, they are generally small compared to total stocks in the ecosystem. Stocks of nutrients remaining in the soil after short-term (1–3 years) cultivation appear to be adequate for tree growth. Trees apparently are able to grow normally in sites that have been cultivated, as long as cultivation does not last for more than 2 or 3 years, as is the case in taungya agriculture.

## References

Aborisade, K.D. and Aweto, A.O. (1990) Effects of exotic tree plantations of teak (*Tectona grandis*) and gmelina (*Gmelina arborea*) on a forest soil in south-western Nigeria. *Soil Use and Management* 6, 43–45.

Ae, N., Arihara, J., Okada, K., Yoshihara, T. and Johansen, C. (1990) Phosphorus uptake by pigeon pea and its role in cropping systems of the Indian subcontinent. *Science* 248, 477–480.

Alexander, T.G. (1987) Taungya and soil management during the establishment phase of forest plantations in Kerala, India. In: Vergara, N.T. and Briones, M.C. (eds) *Agroforestry in the Humid Tropics.* Environment and Policy Institute, East–West Center, Honolulu, Hawaii, pp. 125–131.

Brinkmann, W.L.F. and Nascimento, J.C. (1973) The effect of slash and burn agriculture on plant nutrients in the tertiary region of Central Amazonia. *Acta Amazonica* 3, 55–61.

Denevan, W.M. (1971) Campa subsistence in the Gran Pajonal, eastern Peru. *The Geographical Review* 61, 496–518.

Dubroeucq, D. and Sanchez, V. (1981) *Caracteristicas Ambientales y Edaficas del Area Muestra San Carlos de Rio Negro-Solano.* Ministerio del Ambiente y de los Recursos Naturales Renovables, Serie Informe Cientifico DGSIIA/IC/12. Caracas, Venezuela.

Harris, D.R. (1971) The ecology of swidden cultivation in the upper Orinoco rain forest, Venezuela. *The Geographical Review,* 61, 475–495.

Jordan, C.F. (1982) Amazon rain forests. *American Scientist* 70, 394–401.

Jordan, C.F. (1985) *Nutrient Cycling in Tropical Forest Ecosystems.* Wiley, Chichester.

Jordan, C.F. (1989) *An Amazonian Rain Forest: The Structure and Function of a Nutrient Stressed Ecosystem and the Impact of Slash-and-Burn Agriculture.* Man and the Biosphere Series, Vol. 2. Parthenon. Carnforth, UK.

Lee, D., Han, X.G. and Jordan, C.F. (1990) Soil phosphorus fractions, aluminum, and water retention as affected by microbial activity in an ultisol. *Plant and Soil* 121, 125–136.

Nye, P.H. and Greenland, D.J. (1964) Changes in the soil after clearing tropical forest. *Plant and Soil* 21, 101–112.

Sanchez, P.A. (1976) *Properties and Management of Soils in the Tropics.* Wiley, New York.

Sanchez, P.A., Villachica, J.H. and Bandy, D.E. (1983) Soil fertility dynamics after clearing a tropical rainforest in Peru. *Soil Science Society of American Journal* 47, 1171–1178.

Takahashi, T., Nagahori, K., Mongkolsawat, C. and Losirikul, M. (1983) Run-off and soil loss. In Kyuma, K. and Pairintra, C. (eds) *Shifting Cultivation: An Experiment at Nam Phrom, Northeast Thailand, and its Implications for Upland Farming in the Monsoon Tropics.* Faculty of Agriculture, Kyoto University, Kyoto, Japan, pp. 84–109.

Tulaphitak, T., Pairintra, C. and Kyuma, K. (1983) Soil fertility and tilth. In Kyuma, K. and Pairintra, C. (eds) *Shifting Cultivation: An Experiment at Nam Phrom, Northeast Thailand, and its Implications for Upland Farming in the Monsoon Tropics.* Faculty of Agriculture, Kyoto University, Kyoto, Japan, pp. 63–83.

Verinumbe, I. and Okali, D.U.U. (1985) The effect of coppiced teak regrowth on soil in a teak-maize agroforest. *Forest Ecology and Management* 12, 37–41.

Zinke, P.J., Sabhasri, S. and Kunstadter, P. (1978) Soil fertility aspects of the Lua' forest fallow system of shifting cultivation. In Kunstadter, P., Chapman, E.C. and Sabhasri, S.S. (eds) *Farmers in the Forest.* University of Hawaii, Honolulu, pp. 134–159.

Chapter 6

# Sustainability of Fast-growing Plantation Forests in the Humid Tropics with Particular Reference to Nutrients

L.A. Bruijnzeel

*Faculty of Earth Sciences, Free University, De Boelelaan 1085, 1081 HV Amsterdam, The Netherlands*

In view of the extent of severely eroding, hydrologically and ecologically disrupted lands in the humid tropics, calls for massive reforestation programmes are becoming more frequent (Eckholm, 1976; Postel and Heise, 1988). Also, in response to the rapid increase in population numbers and the associated rising demands for fuel, construction and pulp wood, the area of forest plantations in the tropics almost tripled between 1965 and 1980 and the rate of planting in the 1980s was expected to double that of the 1970s (Evans, 1982).

On the other hand, concern has been expressed about the possible negative effects that fast-growing tree plantations may have upon soil nutrient reserves, especially when grown in short rotations on poor tropical soils (Chijioke, 1980; Hase and Fölster, 1983; Jorgensen and Wells, 1986; Ruhiyat, 1989). Depletion of soil nutrient reserves during the cropping phase of shifting cultivation or after conversion of tropical rain forest to pastures manifests itself in a more or less rapid decline in crop yields and is well documented (Nye and Greenland, 1960; Sanchez, 1976; Buschbacher, 1984; Andriesse, 1987).

This chapter examines what is known of the overall nutrient balances of tropical forest plantations, focusing particularly on nutrient losses during plantation harvest and establishment, with a view to derive conclusions as to their sustainability

## Methodological Aspects

When a natural forest is replaced by a tree plantation, the evaluation of a possible future decrease in productivity (often called the 'second rotation problem') becomes a rather complex affair. To start with, the determination of nutrient losses in harvested produce is a laborious task, surrounded with practical problems at every stage (Auchmoody and Greweling, 1979). In addition, the quantification of nutrient inputs (via bulk precipitation and rock weathering) and losses (soil erosion, deep leaching) requires the long-term collection of climatic and hydrological data as well as the analysis of numerous water samples (Likens *et al.*, 1977). Finally, there is the long time-span inherent to forest production itself.

An alternative approach, therefore, involves the comparison of soil, or ecosystem, nutrient stores in a 'false time series' (i.e. pre-treatment vs. post-treatment situations: Hase and Fölster, 1983; Buschbacher, 1984). However, this method suffers from the (severe) limitation that temporal changes may be confounded by spatial variability in soil characteristics (Bruijnzeel, 1990). In addition, it is rather insensitive and only major differences between sites can be detected in this way (Verstraten, 1980).

Whatever the approach adopted, there is the additional difficulty of interpreting the results. This is caused by the inability to define clearly the soil nutrient store in forestry. As pointed out by Hase and Fölster (1983), the determination of amounts of available nutrients in forest soils by chemical extraction methods still lacks verification by bioassay techniques. Also, the lower boundary of the forest root network is often poorly defined and it becomes difficult to establish to what extent nutrients that are released by weathering will be available for uptake by the trees or leached from the system (Bruijnzeel, 1991).

## Available Data

### Nutrient immobilization

Information on nutrient immobilization in tropical tree plantations is gradually becoming available. Examples are the studies on *Pinus caribaea* and *Gmelina arborea* in Nigeria and Brazil (Egunjobi and Bada, 1979; Chijioke, 1980; Russell, 1983), on *Tectona grandis* in Nigeria (Nwoboshi, 1983) and Venezuela (Hase and Fölster, 1983), on *P. patula* and *Cupressus lusitanica* in Tanzania (Lundgren, 1978), on *P. merkusii* and *Agathis dammara* in Indonesia (Bruijnzeel, 1984; Bruijnzeel and Wiersum, 1985), and on *Eucalyptus* spp. in Brazil (Bellote *et al.*, 1980; Poggiani *et al.*, 1983a,b), in southern India (George, 1984) and in East Kalimantan, Indonesia (Ruhiyat, 1989).

Biomass production of fast-growing hardwood species such as *Gmelina* or eucalypts may (Jorgensen and Wells, 1986) or may not (Russell, 1983; Bruijnzeel, 1984) exceed that of tropical conifers. However, amounts of nutrients incorporated in these hardwood plantations are normally higher than those in softwood plantations of similar age (Russell, 1983; Bruijnzeel, 1984), mainly as a result of the higher nutrient concentrations in hardwood tissue (Jorgensen and Wells, 1986). As such, demands on soil nutrient reserves will differ between species.

**Leaching losses**

In contrast to the growing body of information on tropical tree biomass and nutrient content (and hence on potential nutrient losses upon harvesting) and on nutrient inputs from the atmosphere (Bruijnzeel, 1989), there is a real paucity of reliable information regarding hydrological losses of nutrients from tropical forests and therefore on rates of chemical weathering (Clayton, 1979). Bruijnzeel (1991) critically reviewed the available information and was able to distinguish four different groups of substrates, each with a characteristic hydrological nutrient export pattern.

Calcium and magnesium losses from spodosols and highly leached oxisols were distinctly lower than for the overall ultisol/oxisol group, whereas losses from sites with inceptisols exceeded those from the latter group. Still higher losses were recorded for sites underlain by rocks that were particularly rich in these elements, such as limestones, (calcareous) shales and slates. Potassium exports showed a similarly upward trend over the first three groups with some overlap between groups. Potassium losses from the fourth group were low, reflecting the scarcity of the element in the underlying bedrock.

Even less is known with respect to the critical early phases of plantation establishment (clearing, burning, planting) during which period both volumes and nutrient concentrations of water percolating through the soil are enhanced and losses via leaching (Russell, 1983; Parker, 1985; Bruijnzeel, 1990) and soil erosion (Kang and Lal, 1981; Hudson *et al.*, 1983; Lal, 1983) are likely to be maximal. Studies of this kind are now in progress in Indonesia (cf. Bons, 1990) and Sabah, Malaysia (cf. Malmer, 1990). Meanwhile, it is unfortuante that the only two studies for which results on changes in water chemistry during the conversion of tropical forest to plantations are available (i.e. Jari Florestal, Brazil (Russell, 1983) and Sungai Tekam, Peninsular Malaysia (DID, 1986, 1989)) both suffer from methodological shortcomings. The Brazilian study employed rather arbitrary (albeit not entirely unreasonable) methods to derive (not measure) amounts of drainage associated with various stages of plantation development in a ‘false time series’, whilst results for Sungai Tekam were based on insufficiently frequent sampling (Zulkifli Yusop, 1989). As expected, both studies reported temporarily increased nutrient concentrations in soil- or stream-water (Russell, 1983; DID, 1986, 1989; cf. Buschbacher, 1984; Uhl and Jordan, 1984).

In the absence of reliable estimates of dissolved nutrient exports after clear-cutting tropical forest, it is worthwhile to consider results obtained in warm-temperate areas such as the southeastern USA. (Fischer, 1981; Riekerk, 1983; Hewlett *et al.*, 1984; Van Lear *et al.*, 1985) and eastern Australia (Hopmans *et al.*, 1987). According to Hewlett *et al.* (1984) a 'typical southern clear-cut operation' in mixed secondary forest on ultisols/ ultic alfisols produced only minor and short-lived increases in dissolved ion exports. These were essentially caused by the increase in water yield following the clearcut, since concentrations of individual elements either did not respond to the treatment or decreased because of a dilution effect. Slash was not burned but roller-chopped before the area was planted mechanically with loblolly pine (Hewlett *et al.*, 1984).

Repeated pre-harvest, low-intensity prescribed fires to control understory vegetation at a comparable site in South Carolina had no effect on water quality either (Van Lear *et al.*, 1985). Similarly, clearcutting followed by roller-chopping and replanting of natural pine forest on poorly drained sandy ultisols in coastal Florida resulted in an equally modest effect on hydrological nutrient exports (Riekerk, 1983). Effects on both stream-flow and nutrient losses during the first year after the treatment were roughly two to three times greater when cutting was followed by mechanical stump removal and burning/windrowing of slash (cf. Kang and Lal, 1981). Effects were already negligible during the second year after clearing (Riekerk, 1983). It should be added, however, that the 2-year post-treatment period was noticeably dry, whilst the initial calibration period (1 year) had been quite wet (Riekerk, 1983). Unfortunately, any effects of these climatic irregularities on the comparison of pre- and post-treatment hydrological nutrient losses were not addressed properly in this study. In both treatments, however, the extra losses were less than annual nutrient inputs from atmospheric sources (Riekerk, 1983). In addition, they were much smaller than nutrient losses associated with harvested produce and burning (Fisher, 1981).

Equally small dissolved ion exports were reported by Hopmans *et al.* (1987) following clearcutting and burning of eucalypt forest in southeastern Australia. Apart from somewhat increased potassium concentrations, increases in hydrological nutrient losses during the first year after the operation were entirely due to the increase in water yield. Hopmans *et al.* (1987) ascribed this lack of response to the presence of 30-m-wide buffer strip which had been maintained on either side of the stream channel. Again, hydrological nutrient losses were a mere fraction of the losses due to burning the original forest or of the amounts incorporated in the fast-growing pines that replaced the eucalypts (Stewart *et al.*, 1981; Stewart and Flinn, 1985). Inputs of calcium via rock weathering (as computed via the small catchment mass balance method; Clayton, 1979) were low and might cause productivity problems during future rotations (Hopmans *et al.*, 1987).

It would be premature, however, to conclude from the above examples that leaching losses associated with clearcuts in the humid tropics would also be small. For one thing, the rainfall surplus over evapotranspiration is considerably larger in the tropics and with it the potential intensity of leaching. In addition, due to the large standing biomass of mature tropical forest, considerably larger amounts of nutrients may be removed in harvested stems and burning of logging debris (Hase and Fölster, 1983; Russell, 1983; Stewart and Flinn, 1985; Proctor, 1987). On the other hand, rates of regeneration and nutrient uptake will be faster in the tropics (Uhl and Jordan, 1984; Jaffré, 1985). Clearly, there is a need for more quantitative information in this respect and it makes one look forward to the results of the Malaysian and Indonesian studies mentioned previously.

## Case Studies

No single study has managed adequately to quantify all gains and losses of nutrients for a complete cycle of tropical forest clearing, followed by plantation establishment, maturing and harvesting. However, approximate nutrient budgets for entire forest rotations have been computed for teak plantations on eutrophic soils in Venezuela (Hase, 1981; Hase and Fölster, 1983), for fast-growing conifers on fairly fertile volcanic soils in Indonesia (Bruijnzeel, 1984; Bruijnzeel and Wiersum, 1985) and for fast-growing hard- and softwoods on oligotrophic soils in Amazonia (Russell, 1983; 1987). These are discussed below.

### Teak and the Western Llanos of Venezuela

In western Venezuela sites with light textured soils ('Banco' sites with inceptisols) are considered suitable for reforestation with teak. The natural vegetation of these sites is a semi-evergreen seasonal forest (average rainfall 1800 mm per year with a pronounced dry season), which may attain an above-ground biomass figure of about 400 t $ha^{-1}$ (Hase and Fölster, 1982). Hase and Fölster (1983) presented data on nutrient stocks in vegetation and soils (0–100 cm) of ten teak plantations ranging in age between 6 months and 9 years as well as in mature forest of the type that preceded the plantations. In addition, they measured atmospheric nutrient inputs for 1 year and derived contributions by weathering from the observation that 5% of the primary silicate minerals present in the sand and silt fractions of the soils had been transformed into kaolinitic clays, combining this finding with an estimated age of the top 100 cm of the soil of about 1000 years (Hase, 1981).

A tentative nutrient budget for the first 80 years after forest clearing and planting teak was constructed on the basis of the above original data in combination with data on timber yield of teak plantations elsewhere in the tropics, stemwood nutrient contents of 33- and 38-year-old teak trees in northern India, and averaged leaching losses for a number of temperate and tropical forest ecosystems (Hase and Fölster, 1983). Erosion during plantation establishment or at a later stage (Bell, 1973) was not taken into consideration. Table 6.1 summarizes the results.

Considerable amounts of nutrients appear to be lost from the site with the removal of merchantable timber upon clearing the natural forest and as a result of increased leaching during the first year (Hase, 1981). Still larger amounts are removed in harvested teak boles at the end of the first rotation. Total nutrient losses are in all cases significantly higher than additions via precipitation and weathering, thereby depleting soil nutrient reserves.

As shown in Table 6.1, by far the most serious loss associated with the conversion to teak and the subsequent harvest of logs is that of calcium. Although these estimates must be considered as orders of magnitude only, because of the uncertainty introduced by the liberal use of data from the literature, it is clear that calcium will be the prime element limiting future productivity, possibly so already during the second rotation (Hase and Fölster, 1983). A rough check of the latter assertion can be obtained by assuming similar nutrient inputs and losses (harvest of teak boles and 'steady state' leaching only) for the second rotation and comparing the net losses so obtained with approximate nutrient stocks present at the start of the latter (see footnotes to Table 6.1). Indeed, estimates of calcium exports associated with the second rotation vary between 90 and 395% of initial stocks (Table 6.1). Most of this variation is due to the variability of calcium concentrations in stemwood of mature teak trees as reported in the literature (Hase and Fölster, 1983). However, similarly high concentrations of calcium and magnesium were found in the stems of 25-year-old teak on inceptisols in Indonesia, whilst values for phosphorus and potassium were 19 and 194% higher, respectively, than the ones on which the computations of Table 6.1 are based (Bruijnzeel, 1983a). Interestingly, rates of soil nutrient depletion were much lower for the less fertile sandier Banco sites, reflecting the lower productivity (and rate of nutrient immobilization) of these sites (Hase, 1981).

Since the observations of Hase and Fölster took place in 1977/78, it would be interesting to see whether repeated sampling in the oldest teak stands (which would be about 20 years old now) would confirm the trends predicted in Table 6.1. Also, since the hydrology of the natural forest in the area is well researched (Franco, 1979), additional observations of soil- and

**Table 6.1.** Approximate nutrient budget (kg $ha^{-1}$)[1] for the first rotation of teak production (80 years) on loamy Banco soils in western Venezuela.

| | P | K | Ca | Mg |
|---|---|---|---|---|
| Inputs | | | | |
| Rainfall | 40 | 475 | 580 | 165 |
| Weathering | ? | 425–535 | 50–110 | 105–135 |
| Total input | > 40 | 900–1010 | 630–690 | 270–300 |
| Outputs | | | | |
| Harvest of mature forest | 25 | 430 | 980 | 50 |
| Extra leaching | ? | 0–700 | ? | ? |
| Steady leaching | 10 | 185 | 480 | 170 |
| Harvested boles | 490 | 805 | 2935–4680 | 545 |
| Total losses | 525 | 1770 | 4395–6140 | 765 |
| *Net losses* | *485* | *760–870* | *3705–5510* | *465–495* |
| Total stock in natural forest | 4890[2] | 2290 | 6725 | *1760* |
| *Net loss as % of total stock* | *10*[3] | *33–38* | *55–82* | *26–28* |
| Remaining stock at start of 2nd rotation[4] | 4400 | 1215–1535 | 1145–3025 | 1180–1300 |
| Inputs | > 40 | 900–1010 | 630–690 | 270–300 |
| Outputs | 500 | 990 | 3415–5160 | 715 |
| Net losses | 460 | −20–90 | 2725–4530 | 415–445 |
| *Net loss as % of stock at start 2nd rotation* | *10*[5] | *0–7* | *90–395* | *32–38* |

Adapted from Hase and Fölster, 1983.
[1]Rounded off to nearest 5.
[2]Of which ca. 2300 kg $ha^{-1}$ as available phosphorus in the soil.
[3]21% of available phosphorus reserves.
[4]Excluding slash from teak harvest.
[5]25% of available phosphorus reserves.

stream-water composition would improve estimates of amounts of nutrients supplied by weathering and lost via leaching. However, as pointed out by Hase and Fölster (1983), most of the uncertainty surrounding the long-term nutrient dynamics of these teak plantations is related to stemwood nutrient contents of mature trees and it is here that further information is needed most. No follow-up work has been undertaken since the original study (dr. H. Fölster, pers. commun.).

## Conifers in Central Java, Indonesia

Bruijnzeel and Wiersum (1985) presented a nutrient balance sheet for ideally stocked plantations of *Agathis dammara* grown in rotations of 40 years on moderately fertile andosols in Java. Atmospheric nutrient inputs, contributions by weathering, and deep leaching were determined by means of the 'small-watershed mass balance method' (Bruijnzeel, 1983b). Losses in harvested produce were based on measurements of biomass and nutrient contents in a 'false time series' of four plantations ranging in age between 7 and 35 years and growing on the same soil type (Bruijnzeel, 1983a).

Plantation forestry in Java is usually practised on steep slopes and employs the taungya system during the early stages of plantation establishment. As such, the soil is exposed for about 3 years and during this period soil erosion may occur. Bruijnzeel and Wiersum (1985) estimated nutrient losses associated with erosion during taungya from pre-war erosion studies in Java (Coster, 1938; Genggrijp, 1941) and made a distinction between 'moderate' (1 cm per year) and 'heavy' (2.5 cm per year) erosion. Recent work on similar soils in West and East Java by Bons (1990) and by Rijsdijk and Bruijnzeel (1990) respectively suggested erosion rates of 35–70 t $ha^{-1}$ per year (ca. 0.5–1.1 cm per year) during taungya, the bulk of which was produced by those parts of the fields that were used as access routes (e.g. field boundaries). No estimates of extra losses due to increased leaching during taungya were made at the time (Bruijnzeel and Wiersum, 1985). Their results are summarized in Table 6.2, supplemented with an estimate of initial leaching losses (see footnotes). To facilitate comparison with Table 6.1, nutrient losses associated with trunk removal (including bark) are given, even though the prevailing practice generally involves on-site debarking. Since branches are usually collected for use as fuelwood and (in certain parts of Java) *Agathis* leaves are also used for fodder/compost making, several harvesting regimes have been included in Table 6.2. A comparison of total gains and losses of nutrients over the rotation shows that inputs are sufficient to allow the regular harvesting of bole wood (including bark) plus branches, but that total tree harvesting is most likely to deplete soil reserves to an unacceptable level, especially so for phosphorus (Table 6.2).

**Table 6.2.** Nutrient budget (kg $ha^{-1}$)[1] for an ideally stocked plantation of *Agathis dammara* under various management conditions and a rotation period of 40 years.

| | P | K | Ca | Mg |
|---|---|---|---|---|
| Inputs | | | | |
| Rainfall | 35 | 385 | 395 | 160 |
| Weathering | 200 | 1575 | 3060 | 1565 |
| Total input | 235 | 1960 | 3455 | 1725 |
| Outputs | | | | |
| Leaching | 30 | 880 | 1160 | 1220 |
| Extra leaching during taungya[2] | 2 | 25 | 15 | 15 |
| Erosion[3] | 2 | 15–25 | 20–140 | 10–40 |
| Harvested crops[4] | 20 | 40 | 20 | 7.5 |
| Harvest of stemwood only | 105 | 435 | 660 | 155 |
| Idem + bark | 145 | 560 | 1295 | 250 |
| Idem + branch | 205 | 855 | 1680 | 330 |
| Total tree[5] | 350 | 1655 | 3315 | 705 |
| *Losses/gains* | | | | |
| Bole harvest | 0.85 | 0.78 | 0.74 | 0.88 |
| Bole + branch | 1.10 | 0.93 | 0.86 | 0.92 |
| Total tree | 1.72 | 1.34 | 1.33 | 1.14 |
| Nutrients in soil (0–100 cm)[6] | 90–110 | 540–1340 | 750–4130 | 340–1220 |
| *Net loss as % of reserves*[7] | *155–189* | *49–122* | *25–151* | *20–72* |

Adapted from Bruijnzeel and Wiersum, 1985.
[1]Rounded off to nearest 5.
[2]Assuming an increase in water yield equal to rainfall interception in a mature plantation (ca. 670 mm (Bruijnzeel, 1988); a doubling of K concentrations and no changes in stream-water concentrations of P, Ca, and Mg (C.A. Bons. pers. commun.).
[3]Assuming a rate of 1 cm per year for 3 years of taungya; available nutrients only.
[4]Upland rice and maize, grain only as straw is usually burned on-site (based on concentrations in Ochse *et al.*, 1961).
[5]Not including undergrowth.
[6]Readily available: extractable by $NH_4$-acetate at pH 7 (Ca, Mg, K) or a 2% solution of citric acid (P).
[7]In case of total tree harvest.

It is important, however, to distinguish between fractions of soil phosphorus that are readily available (ortho-P in soil solution), non-available (locked up in (an)organic compounds) and potentially available (organic P) (Sanchez, 1976). The methods employed by Bruijnzeel and Wiersum (1985) extracted readily available and total phosphorus but not organic phosphorus. As such, their estimate of available phosphorus will be highly conservative. Total amounts present in the soil amounted to 5–6000 kg $ha^{-1}$ (Bruijnzeel, 1983a). NUFFIC-UNIBRAW (1984) reported concentrations of available phosphorus (determined by the Bray-II technique which extracts larger amounts than the relatively mild citrate method) for a *dammara* plantation in East Java underlain by a soil that was morphologically and chemically very similar to the soils encountered by Bruijnzeel and Wiersum, which were up to two times higher. However, even if the available phosphorus stock were double the presently quoted value, total tree harvesting would still exhaust available phosphorus reserves after one full rotation. Further work on the availability of phosphorus to trees in these volcanic soils is necessary.

Bruijnzeel (1984) conducted in similar analysis for *Pinus merkusii* plantations in the vicinity of the above *Agathis* stands grown in rotations of 25 years and on more fertile soils. Nutrient uptake rates over 25 years by the (overstocked) pine stands were considerably lower than those for *Agathis* and in fact enabled total tree harvesting without any serious decline in soil fertility.

## Fast-growing pines in Amazonia

The largest single conversion of tropical rain forest to plantations is situated at Jari Florestal in eastern Amazonia. The plantations (mainly *Pinus caribaea* and *Gmelina arborea*) are mostly on infertile sandy oxisols and ultisols (Russell, 1983).

Russell (1983) determined forest biomass and total ecosystem nutrient stocks for a false time series consisting of (i) undisturbed natural forest, (ii) 6-month-old pines, (iii) an 8.5-year-old *Gmelina* stand, (iv) a 9.5-year-old pine stand, and (v) a 6-month-old pine plantation on a site where *Gmelina* had been grown for 8.5 years. All sites were on sandy ultisols with very similar physical characteristics, on the basis of which Russell assumed their original chemical properties to have been the same as well. As such, all differences in soil nutrient reserves between sites were ascribed to the various treatments whilst possible differences in initial site fertility were ignored.

The hazards of this approach become readily clear from the data presented by Russell (1983) on nutrient stocks in natural forest and mature plantations. For example, total stocks of phosphorus and calcium were larger for the *Gmelina* site than in the natural forest. Although these differences are probably not statistically significant due to the large standard errors of the biomass estimates, it cannot be excluded that this finding reflects a real difference in original site quality. Also, soil reserves of calcium and magnesium in the mature pine stand amounted to only 13 and 18%, respectively, of those in the *Gmelina* plantation whereas amounts of the two elements incorporated in tree biomass differed by only 20% between species (with the higher values for the slightly older pines). Of course, one could think of a variety of factors to explain such differences between sites, ranging from differences in original forest biomass or amounts of slash left upon clearing, to contrasting site fertility. Whatever the true reason, it will be clear that there can be no substitute for recurrent inventories of soil chemical characteristics at the same location (Chijicke, 1980).

In addition to the determination of total system nutrient stocks, Russell (1983) collected data on atmospheric nutrient inputs (for 1 year) and on the chemical composition of free-draining soil water in the various treatments. By combining the information on soil water quality with (theoretical) estimates of annual drainage rates, an idea was obtained of leaching losses associated with the various phases of the time series. In view of the above difficulties with the data set for *Gmelina*, the following discussion concentrates on the conversion of natural forest to pine plantations only and avoids a comparison of total nutrient stocks for different sites (Table 6.3).

Net losses of nutrients associated with the first cycle of forest production were considerable, removing 62% (phosphorus) to 83% (magnesium) of the total nutrient store present before the conversion. Up to 45% of the (gross) losses consisted of nutrients removed by the harvesting of timber from the natural forest whereas only a modest proportion (11–17%) was associated with the harvest of pine stems (with the exception of phosphorus). Leaching losses of calcium, magnesium and potassium appeared sizeable as well, although it should be remembered that these estimates may not be too reliable. Since extra leaching losses during the establishment of the second rotation were small (Russell, 1983), net losses associated with the second cycle were only a fraction (12–21%, again with the exception of phosphorus: 54%) of those determined for the first period (Table 6.3).

Both on the basis of the total nutrient inventory for the mature pine stand given by Russell (1983) and the inferred remaining stocks after the first rotation implied by Table 6.3, it would seem as if magnesium would be the prime element limiting future forest productivity, followed by phosphorus, calcium and potassium. The number of rotations after completion of the first cycle that would be possible before available nutrient stocks would be exhausted, would be less than two (magnesium and phosphorus),

**Table 6.3.** Approximate nutrient budget (kg ha$^{-1}$)[1] for first and second rotations of pine production (10 years each) on sandy ultisols, Jari Florestal, Brazil.

| | P | K | Ca | Mg |
|---|---|---|---|---|
| Inputs | | | | |
| Rainfall[2] | 1.5 | 100 | 160 | 35 |
| Weathering[3] | – | – | – | – |
| Total input | > 1.5 | > 100 | > 160 | > 35 |
| Outputs | | | | |
| Harvest of mature forest | 30 | 375 | 520 | 175 |
| Leaching[4] | 1 | 445 | 480 | 365 |
| Pine harvest[5] | 35 | 100 | 205 | 85 |
| Total losses | 66 | 920 | 1205 | 625 |
| *Net losses* | *65* | *820* | *1045* | *590* |
| Total stock in natural forest | 105 | 1055 | 1420 | 710 |
| *Net loss as % of total stock* | *62* | *78* | *74* | *83* |
| Net losses during second rotation[6] | 35 | 95 | 215 | 110 |
| *Net loss as % of remaining stock*[7] | *59* | *27* | *38* | *63* |
| No. of rotations after 1st before exhaustion[8] | 1.7 | 3.7 | 2.6 | 1.6 |

Adapted from Russell, 1983.
[1]Rounded off to nearest 5.
[2]Underestimate since particulate matter not included in samples.
[3]Considered negligible by Russell (1983).
[4]Added losses estimated by Russell for first year after logging and burning to those for young pine plantation ('second' year) and interpolating between the latter and value for mature pine to obtain losses for intermediate years; no extra losses assumed for volatilization.
[5]At age 10, values extrapolated from data for 9.5-year-old stand.
[6]Assuming steady-state leaching (Russell, 1983) and pine bole harvest.
[7]Computed as initial stock minus losses associated with first rotation and adding nutrients contained in branches and foliage of mature pines.
[8]Remaining stocks after 1st cycle divided by net loss of 2nd cycles.

two to three (calcium) and less than four (potassium). Similar figures were derived by Russell himself, with the notable exception of potassium for which he computed 16.7 rotations before complete exhaustion. Since the latter value was based on a comparison of nutrient inventories of contrasting sites, the more recently derived value of 3.7 may be more realistic, also in light of the values obtained for the other elements (Table 6.3).

## Conclusion

It must be concluded, therefore, that (pine) plantation forestry at Jari, and presumably also at other locations experiencing similar soil conditions, is not sustainable without extra inputs from fertilizers (Table 6.3). Fertilizers are, of course, already widely used in extractive tree plantations such as rubber or oil palm (Thajib and Pushparadjah, 1984). An extra problem in the case of fertilized plantations consists of increased leaching losses (Agamuthu and Broughton, 1982; Pushparadjah, 1982; Foong *et al.*, 1983; Chang and Chow, 1985). An evaluation of the economic feasibility of fertilizer use in non-extractive tropical tree plantations (cf. Evans, 1982) is beyond the scope of this analysis but the data presented thus far do indicate the need for fertilizing if poor tropical soils are to sustain intensive plantation forestry (Table 6.3). In addition, they may even be needed on more fertile substrates in the case of whole-tree harvesting (Table 6.2) or when a species requires a certain key nutrient in relatively large quantities (Table 6.1).

Additional information on 'extra' nutrient losses associated with erosion and leaching during plantation establishment and for a variety of substrates would be desirable but the available data suggest these to be (much) smaller generally than losses in harvested produce.

## Acknowledgement

I wish to thank Mr Kees Bons for the information which enabled me to compute the extra leaching loss during 'taungya' as presented in Table 6.2.

## References

Agamuthu, P. and Broughton, W.J. (1982) Nitrogen cycling in a legume-oil palm ecosystem in Malaysia. In Wetselaar, W., Simpson, J.R. and Rosswall, T.R. (eds) *Nitrogen Cycling in South-east Asia Wet Monsoonal Ecosystems.* Australian Academy of Science, Canberra, pp. 113–118.

Andriesse, J.P. (1987) *Monitoring Project of Nutrient Cycling in Soils Used for Shifting Cultivation under Various Climatic Conditions in Asia.* Part I – Text, part II – Annexes. Royal Tropical Institute, Amsterdam, The Netherlands.

Auchmoody, L.R. and Greweling, T. (1979) Problems associated with chemical estimates of biomass. In Leaf, A.L. (ed.) *Proceedings of the Symposium on Impacts of Intensive Harvesting on Forest Nutrient Cycling.* New York State University, Ithaca, New York, pp. 190–210.

Bell, T.I.W. (1973) Erosion in the Trinidad teak plantations. *Commonwealth Forestry Review* 52, 223–233.

Bellote, A.F.J., Sarruge, J.R., Haag, H.P. and De Oliveira, G.D. (1980) Nutrient extraction and exportation by *Eucalyptus grandis* Hill ex Maiden related to age. *IPEF, Piracicaba* 20, 1–23.

Bons, C.A. (1990) Accelerated erosion due to clearcutting of plantation forest and subsequent 'Taungya' cultivation in upland West-Java, Indonesia. *International Association of Hydrological Sciences Publication* 192, 279–288.

Bruijnzeel, L.A. (1983a) *Hydrological and Biogeochemical Aspects of Man-made Forests in South-Central Java, Indonesia.* PhD thesis, Free University, Amsterdam, 256 pp.

Bruijnzeel, L.A. (1983b) The chemical mass balance of a small basin in a wet monsoonal environment and the effect of fast-growing plantation forest. *International Association of Hydrological Sciences Publication* 141, 229–239.

Bruijnzeel, L.A. (1984) Immobilization of nutrients in plantation forests of *Pinus merkusii* and *Agathis dammara* growing on volcanic soils in Central Java, Indonesia. In Thajib, A. and Pushparadjah, E. (eds) *Soils and Nutrition of Perennial Crops.* Malaysian Soil Science Society, Kuala Lumpur, pp. 19–29.

Bruijnzeel, L.A. (1988) Estimates of evaporation in plantations of *Agathis dammara* Warb, in south-central Java, Indonesia. *Journal of Tropical Forest Science* 1, 145–161.

Bruijnzeel, L.A. (1989) Nutrient content of bulk precipitation in south-central Java, Indonesia. *Journal of Tropical Ecology* 5, 187–202.

Bruijnzeel, L.A. (1990) *Hydrology of Moist Tropical Forests and Effects of Conversion: a State-of-Knowledge Review.* Humid Tropics Programme of IHP-UNESCO, Paris, 224 pp.

Bruijnzeel, L.A. (1991) Nutrient input–output budgets of moist tropical forest ecosystems: a review. *Journal of Tropical Ecology* 7, 1–24.

Bruijnzeel, L.A. and Wiersum, K.F. (1985) A nutrient balance sheet for *Agathis dammara* Warb. plantation forest under various management conditions in central Java, Indonesia. *Forest Ecology and Management* 10, 195–208.

Buschbacher, R.J. (1984) *Changes in Productivity and Nutrient Cycling Following Conversion of Amazonian Rain Forest to Pasture.* PhD thesis, University of Georgia, Athens, Ga. 193 pp.

Chang, K.C. and Chow, C.S. (1985) Some questions of the leaching losses of soil nutrients. *Oléagineux* 40, 233–240.

Chijicke, E.O. (1980) Impact on Soils of Fast-growing Species in Lowland Humid Tropics. *FAO Forestry Paper* no. 21, FAO, Rome, 111 pp.

Clayton, J.L. (1979) Nutrient supply to soil by rock weathering. In Leaf, A.L. (ed.) *Proceedings of the Symposium on Impacts of Intensive Harvesting on Forest Nutrient Cycling.* New York State University, Ithaca, New York, pp. 75–96.

Coster, C. (1938) Superficial runoff and erosion in Java. *Tectona* 31, 613–728 (in Dutch, with English summary).

DID (1986) Sungai Tekam Experimental Basin. Transition Report July 1980 to June 1983. *Water Resources Publication* No. 16. Drainage and Irrigation Department, Ministry of Agriculture, Kuala Lumpur, Malaysia, 102 pp.

DID (1989) Sungai Tekam Experimental Basin. Final Report July 1977 to June 1986. *Water Resources Publication* no. 20. Drainage and Irrigation Department, Ministry of Agriculture, Kuala Lumpur, Malaysia, 107 pp.

Eckholm, E. (1976) *Losing Ground.* W.W. Norton, New York, 223 pp.

Egunjobi, J.K. and Bada, S.O. (1979) Biomass and nutrient distribution in stands of *Pinus caribaea* L. in the dry forest zone of Nigeria. *Biotropica* 11, 130–135.

Evans, J.C. (1982) *Plantation Forestry in the Tropics.* Clarendon Press, Oxford, 452 pp.

Fischer, R.F. (1981) Impact of intensive silviculture on soil and water quality in a coastal lowland. In Lal, R. and Russell, E.W. (eds) *Tropical Agricultural Hydrology.* J. Wiley, New York, pp. 299–309.

Foong, S.F., Syed Sofi, S.O. and Tan, P.Y. (1983) A lysimetric simulation of leaching losses from an oil palm field. In Malaysian Soil Science Society. *Proceedings of the Seminar on Fertilizers in Malaysian Agriculture.* Malaysian Society of Soil Science, Kuala Lumpur, pp. 45–68.

Franco, W. (1979) Die Wasserdynamik einiger Waldstandorte der West-Llanos Venezuelas und ihre Beziehung zur Saisonalität des Laubfalls. *Göttinger Bodenkundliche Berichte* 61, 1–201.

George, M. (1984) Nutrients removal from *Eucalyptus* plantations by harvesting wood. *Indian Forester* 110, 61–67.

Gonggrijp, L. (1941) Erosion experiments. *Tectona* 34/35, 200–220 (in Dutch, with English summary).

Hase, H. (1981) Nährstoffreserven auf Banco-Standorten der Waldreserve Caparo/Venezuela unter besonderer Berücksichtigung der Plantagenwirtschaft mit Teak (*Tectona grandis*). *Göttinger Bodenkundliche Berichte* 66, 1–152.

Hase, H. and Fölster, R.H. (1982) Bioelement inventory of a tropical (semi)-evergreen seasonal forest on eutrophic alluvial soils, Western Llanos, Venezuela. *Oecologia Plantarum* 17, 331–346.

Hase, H. and Fölster, R.H. (1983) Impact of plantation forestry with teak (*Tectona grandis*) on the nutrient status of young alluvial soil in west Venezuela. *Forest Ecology and Management* 6, 33–57.

Hewlett, J.D., Post, H.E. and Doss, R. (1984) Effect of clear-cut silviculture on dissolved ion export and water yield in the Piedmont. *Water Resources Research* 20, 1030–1038.

Hopmans, P., Flinn, D.W. and Farrell, P.W. (1987) Nutrient dynamics of forested catchments in southeastern Australia and changes in water quality and nutrient exports following clearing. *Forest Ecology and Management* 20, 209–231.

Hudson, J., Kellman, M., Sanmugadas, K. and Alvarado, C. (1983) Prescribed burning of *Pinus oocarpa* in Honduras. I. Effects on surface runoff and sediment loss. *Forest Ecology and Management* 5, 269–281.

Jaffré, T. (1985) Composition minérale et stocks de bioéléments dans la biomasse épigée de recrûs forestiers en Côte d'Ivoire. *Acta oecologica* 6/20, 233–246.

Jorgensen, J.R. and Wells, C.G. (1986) Tree nutrition and fast-growing plantations in developing countries *The International Tree Crops Journal* 3, 225–244.

Kang, B.T. and Lal, R. (1981) Nutrient losses in water runoff from agricultural catchments. In Lal, R. and Russell, E.W. (eds) *Tropical Agricultural Hydrology.* J. Wiley, New York, pp. 153–161.

Lal, R. (1983) Soil erosion in the humid tropics with particular reference to agricultural land development and soil management. *International Association of Hydrological Sciences Publication* 140, 221–239.

Likens, G.E., Bormann, F.H., Pierce, R.S., Easton, J.S. and Johnson, N.M. (1977) *Biogeochemistry of a Forested Ecosystem.* Springer Verlag, New York, 146 pp.

Lundgren, B. (1978) Soil conditions and nutrient cycling under natural and plantation forests in Tanzanian highlands. *Reports in Forest Ecology and Forest Soils* no. 31. Swedish University of Agricultural Science, Uppsala, 426 pp.

Malmer, A. (1990) Stream suspended sediment load after clear-felling and different forestry treatments in tropical rainforest, Sabah, Malaysia. *International Association of Hydrological Sciences Publication* 192, 62–71.

NUFFIC-UNIBRAW (1984) *Soils and Soil Conditions, Kali Konto Upper Watershed, East Java. Main Report.* NUFFIC-UNIBRAW Soil Science Project, Universitas Brawijaya and Proyek Kali Konto, Malang, Indonesia.

Nwoboshi, L.C. (1983) Growth and nutrient requirements in a teak plantation age series in Nigeria. *Forest Science* 29, 159–165.

Nye, P.H. and Greenland, D.J. (1960) The soil under shifting cultivation, *Technical Communication* no. 51, Commonwealth Bureau of Soils, Harpenden, UK, 152 pp.

Ochse, J.J., Soule, M.J., Dijkman, M.J. and Wehlburg, C. (1961) *Tropical and Subtropical Agriculture.* MacMillan, New York.

Parker, G.G. (1985) *The Effect of Disturbance on Water and Solute Budgets of Hillslope Tropical Rainforest in Northeastern Costa Rica.* PhD thesis, University of Georgia, Athens, Ga., USA, 161 pp.

Poggiani, F., Couto, H.T.Z. and Suiter Filho, W. (1983a) Biomass and nutrient estimates in short-rotation intensively cultured plantation of *Eucalyptus grandis. IPEF, Piracicaba* 23: 37–42 (in Portuguese).

Poggiani, F., Couto, H.T.Z., Corradini, L. and Fazzio, E.C.M. (1983b) Exportation of biomass and nutrients associated with the removal of trunks and branches from plantations of *Eucalyptus saligna. IPEF, Piracicaba* 25, 37–39 (in Portuguese).

Postel, S. and Heise, L. (1988) Reforesting the Earth. *Worldwatch paper* no. 83. Worldwatch Institute, Washington.

Proctor, J. (1987) Nutrient cycling in primary and old secondary rain forests. *Applied Geography* 7, 135–152.

Pushparadjah, E. (1982) Nitrogen cycle in rubber (*Hevea*) cultivation. In Wetselaar, W., Simpson, J.R. and Rosswall, T. (eds), *Nitrogen Cycling in South-east Asian West Monsoonal Ecosystems.* Australian Academy of Science, Canberra, pp. 101–108.

Riekerk, H. (1983) Impacts of silviculture on flatwoods runoff, water quality, and nutrient budgets. *Water Resources Bulletin* 19, 73–79.

Rijsdijk, A. and Bruijnzeel, L.A. (1990) Erosion, sediment yield and land-use patterns in the upper Konto watershed, East Java. Indonesia. *Konto River Project Communication* no. 18, 2 Volumes. Konto River Project, Malang, Indonesia, 58 and 150 pp.

Ruhiyat, D. (1989) Die Entwicklung der standörtlichen Nährstoffvorräte bei naturnaher Waldewirtschaftung und im Plantagenbetrieb, Ostkalimantan. *Göttinger Beiträge zur Land- und Forstwirtschaft in den Tropen und Subtropen* no. 35: 206 pp.

Russell, C.E. (1983) *Nutrient Cycling and Productivity in Native and Plantation Forests at Jari Florestal, Para, Brazil.* PhD thesis, University of Georgia, Athens, Ga., USA. 130 pp.

Russell, C.E. (1987) Plantation forestry. Case study no. 9: the Jari Project, Para, Brazil. In: Jordan, C.F. (ed.) *Amazonian Rain Forests. Ecosystem Disturbance and Recovery.* Springer, New York, pp. 76–89.

Sanchez, P.A. (1976) *Properties and Management of Soils in the Tropics.* J. Wiley, New York, 618 pp.

Stewart, H.T.L. and Flinn, D.W. (1985) Nutrient losses from broadcast burning of *Eucalyptus regnans* debris in Northeast Victoria. *Australian Forest Research* 15, 321–332.

Stewart, H.T.L., Flinn, D.W. and James, J.M. (1981) Biomass and nutrient distribution in radiata pine. In Turvey, N.D. *et al.* (eds) *Productivity in Perpetuity.* CSIRO, Canberra A.C.T., Australia, pp. 173–185.

Thajib, A. and Pushparadjah, E. (1984) *Soils and Nutrition of Perennial Crops.* Malaysian Soil Science Society, Kuala Lumpur, 453 pp.

Uhl, C. and Jordan, C.F. (1984) Succession and nutrient dynamics following forest cutting and burning in Amazonia. *Ecology* 65, 1476–1490.

Van Lear, D.H., Douglass, J.E., Cox, S.K. and Augspurger, M.K. (1985) Sediment and nutrient export in runoff from burned and harvested pine watersheds in the South Carolina Piedmont. *Journal of Environmental Quality* 14, 169–174.

Verstraten, J.M. (1980) Water–rock interactions. A case study in a very low-grade metamorphic shale catchment in the Ardennes, NW Luxembourg. *BGRG Research Monograph No. 2.* Geobooks, Norwich, UK, 244 pp.

Zulkifli Yusop (1989) Effects of selective logging methods on dissolved nutrient exports in Berembun watershed, Peninsular Malaysia. Paper presented at the *FRIM-IHP-UNESCO Regional Seminar on Tropical Forest Hydrology,* Kuala Lumpur, 4–9 September 1989, 26 pp.

# Chapter 7
# Theoretical Basis for Taungya and its Improvement

Jiragorn Gajaseni[1] and Carl F. Jordan[2]

[1] *Department of Biology, Faculty of Science, Chulalongkorn University, Bangkok 10500, Thailand.*
[2] *Institute of Ecology, University of Georgia, Athens, 30602, Georgia, USA*

---

The concept of 'limiting factors' has played a central role in agricultural research, since plants will compete for limiting factors such as light, nutrients or water. Important research topics have included the amount of nutrients and water that must be given to crops to overcome limiting factors. There has been, until recent years, relatively little research on the cropping system itself. A farmer or researcher selected a crop that was important economically, or for some other reason, and then ploughed the field and produced a monoculture of that crop. Frequently, the only question was, by what methods could productivity of that crop in that field be maximized.

Recently, goals besides maximizing production have emerged. With increasing prices of fuel and fertilizers and pesticides, there has been an interest in farming systems with lower input requirements. With increases in agricultural problems such as nutrient contamination of ground water, erosion, pesticide runoff, irrigation and salinization, interest has begun in alternative approaches to management of agricultural systems (National Research Council, 1989). In these approaches, other factors such as environmental quality and sustainability are considered, in addition to productivity.

There has been particular interest in alternative cropping systems in countries where capital available for inputs is low, and problems of erosion, pollution, and sustainability are high. Agricultural scientists, many of them working with development agencies, have found that multiple cropping systems are often effective where fuel, fertilizer and pesticide are scarce or unavailable, and where economic and social considerations must be taken into account when agricultural systems are being planned. Part of the benefits of multiple cropping systems comes about because of *positive* interactions between the intercropped species.

A classic example of a multispecies system with positive interactions between species is the association between shade trees with nitrogen-fixing symbioses, and understory coffee trees. The beneficial aspects of this association have long been realized. Beer (1987) has listed some of the possible advantages of including shade trees with perennial crops as follows:

1. stabilizes output, and permits more efficient use of labour and machinery;
2. suppresses weeds;
3. product diversification;
4. control of crop phenology;
5. shade can improve the quality of the coffee;
6. reduces evapotranspiration;
7. lowers excess soil moisture;
8. increases moisture input;
9. extends productive life of the crop;
10. reduces temperature extremes;
11. reduces rain damage;
12. reduces some diseases;
13. reduces wind;
14. improves drainage;
15. provides mulch;
16. reduces erosion;
17. reduces decomposition;
18. recycles nutrients;
19. fixes nitrogen;
20. decreases need for chemicals.

Not all advantages occur in all circumstances. Nevertheless, there are enough advantages, especially in regions where capital is scarce, that the advantages of intercropping frequently outweigh the disadvantage of lower per-hectare yields of crop species.

There is now a great deal of literature on agroforestry, some of which was mentioned in the introduction to this book. Much of it draws attention to the idea that plant–plant interactions need not always be negative and may often be mutually beneficial (Gliessman, 1986). Even if the interactions are competitive, intercropped species may outperform species grown in monoculture.

Two core principles underlie these phenomena:

**1.** When one species has an effect on the environment that causes a negative response in the other species, yet both can utilize necessary resources more efficiently when living together than when living separately, we have the *competitive production principle*, or the *interference production principle.*

**2.** When the environment of one species is modified in a positive way by a second species, such that the first is facilitated by the second, we have the *facilitative production principle*, or simple *facilitation* (Vandermeer, 1990).

The phenomena where the production of component crops in an intercrop is higher than the sum of appropriate monoculture crops has been termed 'overyielding' (Francis, 1986). There has been an increasing amount of effort dedicated to the theoretical basis of overyielding in intercropped systems.

An examination of the agroforestry literature suggests that application of intercropping theory to problems of modern taungya could lead to improvements in the taungya system. For this book on taungya forestry, it would seem important to address the question of whether more complex combinations of trees and trees, and trees and crops could help overcome one of the problems of taungya: the lack of food production or economic income between the time when cropping is stopped and the time when the trees are harvested. For this reason, we have written this chapter to examine how the theory of multiple cropping, or the 'ecology of intercropping' as it has been called, can be applied to taungya.

We first review the theoretical ecology of intercropping, as it has been developed for agroforestry. Then the theory will be applied to taungya.

## Intercropping

### Terminology

*Monoculture* is the term applied to farming where one crop and only that crop is planted in the same field year after year, with the field lying barren during the cold or dry season. For all other approaches, a variety of terminology has been applied. The following scheme described by Andrews and Kassam (1976) has been widely used.

*Multiple cropping* is the general term that contrasts with monoculture, and means growing two or more crops on the same field in a year. When one crop is followed by another in sequence on the same plot, this is termed *sequential cropping. Intercropping* is growing two or more crops simultaneously on the same field.

*Mixed intercropping* is growing two or more crops simultaneously with no distinct row arrangement, as in slash and burn agriculture, or in tropical home gardens. *Row intercropping, strip intercropping,* or *alley cropping* is growing two or more crops where one or more of the crops are planted in rows or strips. One of the 'crops' may be trees which are grown to supply mulch or nitrogen to grain crops. *Relay intercropping* is growing two or more crops simultaneously during part of the life cycle of each.

*Agroforestry* is sometimes used in a very general sense to apply to any multiple cropping system, where crops include trees, livestock, and fish, as well as fruit and grain crops. However, only when the multiple cropping system includes trees can the system be correctly termed 'agroforestry'. In an agroforestry system, the trees can be cultivated for fruit production, to improve the light or nutrient status of the grain crops, or for wood production.

Taungya is a type of agroforestry. It is a practice organized by foresters for the purpose of regenerating forests, or establishing tree plantations. In taungya systems, peasant farmers are allowed to plant crops within taungya plantations for a few years, for social and economic reasons.

The relative advantage or disadvantage of a multiple cropping system compared to a monoculture is gauged by a measure called Land Equivalent Ratio (LER) and Relative Value Total (RVT). The concepts are best explained by example. Let us say that an intercropped hectare of corn and beans produces 10 biomass units of corn and 6 units of beans for a total of 16 units per hectare. Then let us say that a half hectare monoculture of corn produces 6 units of corn and an adjacent half hectare plot produces 4 units of beans for a total of 10 units per hectare. The LER of the intercropped system is $16/10 = 1.6$. The RVT is calculated in the same way, except the units are of economic income instead of biomass production. When the LER is greater than 1.0, the result is sometimes termed 'overyielding'. The LER in an intercropping system can be greater or less than 1.0 depending on the relative strengths of the interactions between species, and upon the 'breadth' of the 'niches' of the species in the intercropping system.

In the following sections, the underlying mechanisms of interactions and niche separation in intercropped systems are summarized. A more detailed analyses has been presented by Vandermeer (1989) in his book entitled *The Ecology of Intercropping.*

## Competition

The dominant paradigm in traditional agronomic and forestry research has been competition. Competition is the factor which determines how much space should be left between individuals in a monocultural plantation. Research on spacing has played an important part in agricultural and forestry research.

Spacing always involves a compromise. On the one hand, the farmer or forester wants to plant as many individuals as possible per hectare. On the other hand, sometime during the development of the individual plant, it will become big enough to compete with its neighbour for some resource, be it light, water, or nutrients. Planting individuals too closely results in competition which can stunt the growth of the whole stand and reduce overall yield below the amount that could be gained by a less dense spacing. Obtaining maximum yield from a field under a given farming system necessitates a planting density that is not too close so that competition inhibits development, nor too far so that there is under-utilization of resources.

Because annual plants are smaller at time of harvest than perennials, spacing for annual crops such as corn is much closer than spacing for trees in a timber plantation. Spacing for maximum yield of annual crops is relatively easy to determine empirically and theoretically.

Optimal spacing for trees in a monoculture plantation is much more difficult to determine. In forestry systems, species are often planted closely together to give good shape to the trees and to suppress weeds, and to provide greater choice when selecting final crop trees. Trees are often planted at distances somewhere between 2 and 4 metres apart. At this distance, there are wasted resources between the trees for the first several years, but often by the third or fourth year, the canopy closes, and there is an optimal use of resources. However, shortly after that, competition begins between the individuals. The canopies of all the trees are at the same height and have the same shape. Leaves of adjacent individuals intersect and overlap, and consequently there is competition for light. The roots are all at the same depth and same stage of development, so there can be competition for nutrients and water.

In some plantations, thinning is carried out to reduce competition. For long-lived timber species, several thinnings may be required during an entire rotation. Thinnings depend not only on biological interactions, but also on economic considerations.

## Facilitation

Facilitation is a process in which two individual plants or two populations of plants interact in such a way that at least one exerts a positive effect on the other. If both individuals or populations exert a positive effect, the process is called *mutualism.*

A classic example of facilitation occurs when corn and beans are intercropped under conditions where application of fertilizers is not economically feasible. Under such conditions, the total yield from one hectare of intercropped corn and beans can be greater than the total yield from a half hectare monoculture of corn plus a half hectare monoculture of beans. The overyielding effect may be due to nitrogen limitation on corn grown in monoculture, and nitrogen fixation by root symbionts of the beans in excess of that required by the beans in the intercrop system.

Interactions between crops are not mutually exclusive. Facilitation and competition can occur simultaneously. The above intercropping example of corn and beans demonstrated facilitation as evidenced by overyielding. However, under those same conditions, a hectare of intercropped corn might have lower yield than a hectare monoculture of corn, indicating competition as well as facilitation exists between corn and beans.

Not all multiple cropping systems, are productively superior to monocultures. However, some combinations are practised because of economic factors, even though they may be productively inferior. For example, open grown coffee may have higher production than shade grown coffee, but the higher rate may be due to the open grown coffee being fertilized, while the shade grown coffee is not. The decision to use commercial fertilizers may depend on the cost of fertilizers. Cultural and consumer factors also may play a role. Slower-grown shade coffee may command higher prices in some markets, so the higher price per unit must be weighed against the lower unit production per hectare of land.

Mycorrhizae may play a role in facilitation. Cecilio and Janos (1987) found that the intercrop advantage of a maize–solanum intercrop was lost in a mycorrhizae-free environment.

Growth of an annual grain crop may be facilitated by the interaction of a woody perennial. The woody perennial produces mulch (leaf and twig litter) which benefits the grain, through suppression of weeds and conservation of soil moisture. Present-day use of such systems in China is described in Chapter 14.

An important mechanism through which mulch from trees benefits interplanted crops is through organic acids released from the organic matter during decomposition, or by the decomposers themselves (Jordan, 1989). Phosphorus deficiency is common in tropical oxisols and ultisols, due to fixation of phosphorus by iron and aluminum present in high concentrations in these soils. However, organic acids such as malic and citric acids can chelate the iron and aluminium, thus leaving more dissolved phosphorous available for uptake by plants (Han, 1989). Piscidic acid, produced in the roots of pigeon pea (*Cajanus cajan* L.) has the capability to solubilize phosphorus bound with iron (Ae *et al.*, 1990).

In greenhouse experiments, Guedes and Jordan (unpublished data) have found that sorghum grown together with pigeon pea in soil low in available phosphorus but high in iron-bound phosphorus grows significantly faster than sorghum grown alone, suggesting that piscidic acid is solubilizing the iron-bound phosphorus in the soil. The facilitation of sorghum growth by pigeon pea occurred only when the roots were in close proximity.

Facilitation also occurs when intercropping reduces herbivory. Vandermeer (1989) has classified such interactions into three categories.

**1.** *The disruptive-crop interaction.* In this interaction, a second species disrupts the ability of a pest to efficiently attack its host. This occurs when the herbivore is less likely to find its host plant because of chemical or physical confusion by the second species, or having found a patch of host plants, is more likely to leave that patch because of encounters with non-host plants that are disagreeable to the herbivore.

**2.** *The trap–crop interaction.* Here, a second species attracts a pest that would normally be detrimental to the principal species. For example, corn, when planted in strips in cotton fields, may attract the cotton bollworm away from the cotton (Lincoln and Isley, 1947) although not always (Vandermeer, 1989).

**3.** *The enemies interaction.* Here, the intercropping situation attracts more predators and parasites on the herbivores, thus reducing these pests through predation and parasitism. Hansen (1983) demonstrated increases in abundances of several predator species in an intercrop system of maize and cowpea in Mexico, suggesting an explanation for the overyielding of that system.

## Niche separation

Every species has its own unique requirement for nutrients, light, and water for optimum growth. Some species may require high levels of nitrogen, but can survive with low amounts of calcium. Other species may require lots of calcium, but can survive with low levels of phosphorus. Some species do well in the shady understory, others demand full sunlight. Some species demand much water, others get by on very little.

No species, when growing in an even-aged monoculture, uses all the available resources; some resources will be 'wasted'. If the crop does not require high levels of potassium, some of it may be leached instead of taken up. If the plant does not require a lot of water, some of the soil moisture may be evaporated instead of transpired. If the plant demands full sunlight for its leaves, a certain amount of light will reach the soil surface unutilized. It can be argued that one reason that uneven-aged, diverse forests are much more common in nature than even-aged single species forests is that there are many 'wasted' resources in the latter type of forests, and in nature, species have evolved to take advantage of unexploited resources, or to 'fill vacant niches' as the phenomena is sometimes described.

If two species have similar niches, that is, similar requirements for nutrients, light and water, they are likely to compete if growing in proximity. It can be hypothesized that associations of plants in nature are organized so as to minimize competition, although this hypothesis is difficult to prove, since competition is hard to measure. Nevertheless, the hypothesis is very compelling.

When overyielding is observed, it can be interpreted as occurring because the niches of the intercropped species are very different, and the available resources are being used more efficiently than they would be by only one species. For example, overyielding of intercrops of legumes and non-legumes can be attributed in part to the very different requirements of the two species for inorganic soil nitrogen (Snaydon and Harris, 1979).

The light resource may be partitioned as a result of the type and angle of leaves of component crops in an intercropped system (Trenbath, 1979). Corn, beans and squash may grow well together because of partitioning of the light environment as well as the soil nutrient resource. Corn is in the grass family, and has leaves which are primarily vertical. During mid-day hours when light enters from above, a proportion of the light may reflect from the leaves of the corn, and be intercepted by the horizontal leaves of the squash. The growth of cocoa in the understory of canopy trees that can be used for timber also illustrates partitioning of the light resource.

Where factors of production are relatively cheap, intercropping may not be desirable. For example, if nitrogen is very cheap relative to the price paid for coffee, it makes economic sense for the farmer to plant coffee in full sunlight and fertilize. However, where factors of production such as nutrients are scarce and expensive, the principal of niche separation suggests that intercropping may be a more economically realistic solution for the farmer. If the cost of nitrogen is relatively expensive, having a shade canopy of nitrogen-fixing trees could be a better strategy.

In developed regions, fertilizers are often relatively cheap, but intercropping may be desirable for other reasons, such as pollution control and erosion control. Monocultures often do not use all the added fertilizers, nor do they completely cover the soil for an entire year. Additional crops could lessen the problems caused by excess fertilizers and bare soil.

## Interactions in Forests

Competition, facilitation and niche separation occur in forest ecosystems as well as in agricultural systems. Separation of niches occurs between species in forests, as well as between agricultural crop species. Little work has been done on quantifying niches of forest trees experimentally, because of the long time interval that would be required. However, ordination of tree species along environmental gradients strongly suggests that each tree species has a unique environmental niche (Whittaker, 1975).

Niche separation need not occur only between species. It also can occur between individuals of the same species, if the individuals have different structures due to different ages. For example, some species of trees in the tropical rain forest can survive and make slow growth in complete shade, but require full sunlight for rapid growth and sexual maturity (Oldeman and vanDijk, 1991). In such forests, yield over a long period of time could be relatively high and sustainable if mature trees but not young trees were harvested. When mature trees were cut, there would be already well established saplings ready to capitalize on the resources newly available after harvest of a nearby large tree.

While there is much interest in management of mixed-species, uneven-aged natural forests (Gomez-Pompa *et al.*, 1991), there has been relatively little effort invested in it, mainly because it still is cheaper and easier to clearcut existing old growth of native hardwoods than it is to establish and manage new stands (Jordan, 1987).

Most forest management is of even-aged monocultures of species or strains that respond rapidly to infusions of fertilizers and to cultivation and elimination of weeds. For production forestry, many foresters prefer monocultures of species such as *Eucalyptus* and *Acacia* that respond better than native species to cultural inputs. Even-aged, evenly-spaced plantations also are often preferred by production foresters, because these type of forests are easier to manage with machinery (Evans, 1982).

Although even-aged monocultures may be more productive under conditions where fertilization and weeding are part of the normal regime, such as the Jari plantation in Brazil (Jordan and Russell, 1989), these procedures may not always be possible. Where they are not, native species may be more desirable because they are better competitors.

To compare performance of exotic monocultures and native species where plantations are initially fertilized and weeded for a few years, we have begun a long-term experiment in the Brazilian Amazon (Batmanian, 1990). The exotic monocultures responded more rapidly than the native species to fertilization and weeding, and the growth during the first 4 years of the plantations was higher by the exotics. However, the most recent survey done in 1991 indicates an increasing advantage by the native species (Batmanian and Jordan, unpublished data). The monocultures are beginning to crowd each other, and growth will stagnate unless thinning takes place. Meanwhile, the native species seem better able to partition the light and the soil environment. The mixed native species have not only a diversity of species, but also have a diversity of structure which adds to their ability to achieve niche separation. We hypothesize that if this study is carried out for long enough, the mixed native species eventually will begin to grow better than the monocultures of the exotics.

As in intercropped agricultural systems, the benefits of diversification in forest systems are greater when the costs of production factors are relatively low compared to the price of the crop or forest product. Thus diversified systems might seem to be more economically appealling in developing countries. Since taungya forest plantations occur almost exclusively in developing countries, increasing species diversity might be a reasonable approach. It would help to overcome the problem of income from the land between the time when cropping of grains is stopped (after the first years) and that when the trees themselves produce income.

## Application of Intercropping Theory to Taungya

A consideration of intercropping theory suggests that more diverse plantations would be better ecologically, as well as providing a more continuous economic income. To test these ideas, we began an experimental approach to improving reforestation by taungya. The experiment is taking place within the context of the Forest Village System of Thailand, described in Chapter 8. At our site near Lampang in the Mae Moh valley of northern Thailand, the plantations are of teak. For the first 2 years, rice and sometimes corn is interplanted amongst the teak seedlings.

Many of the ecological, economic, and social advantages and disadvantages of taungya systems described in Chapter 3 occur at this site. Socially the system is desirable because it reduces the problem of shifting cultivation and provides employment for the peasants. However, it does not solve the problem of the peasants' desire to own land.

Economically the system is less than completely satisfactory, since many peasants suffer economic hardships even though they are part of the system. They need a parcel of land which can be cultivated for longer than 2 years to provide a sustained economic income without the necessity of moving to another parcel of land. Longer tenure at a site would also ensure better stewardship of the site. Ecologically the system is better than simple plantations, since there is better niche utilization during the first few years. However, the severe competition beginning around 8 years is an important drawback to the system.

In an attempt to devise a system that would overcome the latter two problems, we took the ideas of niche separation and facilitation into consideration. We designed a system as follows (Gajaseni and Jordan, 1990) that has a greater diversity of species and of structure, and that has a longer period of productivity of food crops.

In the taungya system of the Forest Village System near Lampang, teak for the last 20 years has been planted at intervals of 4 × 4 m. In our experimental taungya system, we increased the interval to 8 × 8 m. Between each teak, one individual of tamarind, *Tamarindus indica*, a legume, was planted. Then, for the first 2 years, upland rice, squash, pumpkin, and other crops were planted between the tree seedlings as is the custom in this taungya system.

There are several potential advantages to interplanting the tamarind between the teak.

**1.** *Economic advantages.* The tamarind produces a fruit which is highly prized in Thailand, and which has a good economic value. Production of the tamarind fruit should begin at about the time that cultivation of rice is abandoned. Income from sale of the tamarind fruit can extend the economic income for peasants from a particular plot, perhaps until 15 years.

**2.** *Ecological advantages.* The tamarind occupies a different niche than the teak, and thus competition should be much less, and stagnation may not occur. The tamarind will exploit different nutrients at different intensities, and may even make certain nutrients such as nitrogen more readily available to teak. The structural diversity will prevent interlocking of canopy branches and competition for light which often causes stagnation in monocultures of teak. Presumably such competition also will be reduced in the root zone. At the end of approximately 15 years, the canopy of the teak should close and overtop the tamarind, thereby shading it and gradually killing it. Expensive thinning may not be necessary. For the first 15 years of the plantation, the tamarind will provide the side-shading needed to maintain the good form of teak.

Interplanting of the tamarind with the teak is only one of the combinations that we are trying as part of this experiment. A diagram of the field layout (Fig. 7.1) illustrates some of the other combinations.

As part of the experiment, we are also trying uneven-aged management of teak stands. Uneven-aged stands mimic more closely the natural stands of teak. When a mature tree is harvested, a seedling can be planted in the gap, to take advantage of the resources newly made available. With uneven-aged stands, periodic harvests of between 5 and 10 years are possible, in place of the 30–60 year intervals planned for the even-aged stands.

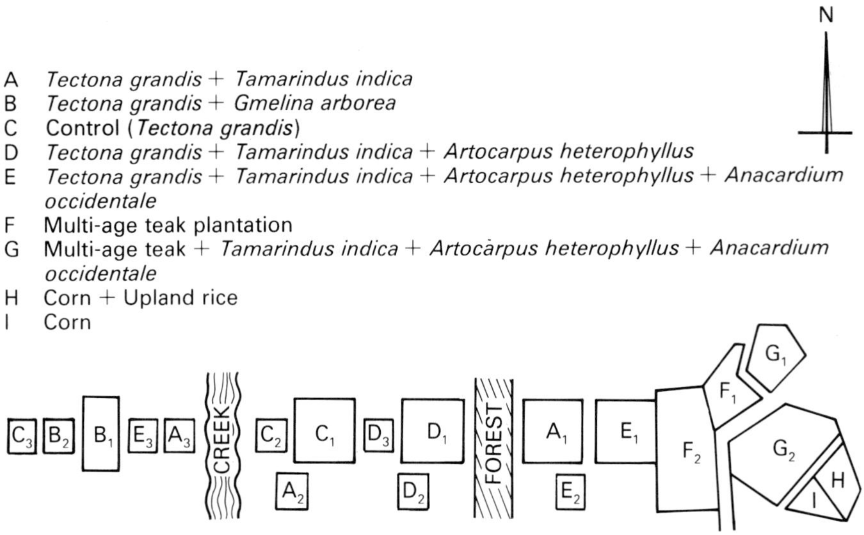


Scale 1: 4500

**Fig. 7.1** Map of the experimental site at the Forest Village taungya site, Mae Moh valley, Thailand.

While theoretically the system should be an improvement, we do not yet have solid evidence. Because the experiment is still less than 3 years old, it is not possible to evaluate its success. However, we plan to continue monitoring the site for several decades, and we hope that similar experiments can be initiated.

## Discussion

There are strong reasons for even-aged monocultures in industrial, temperate countries. Monocultures may occur naturally. Even-aged monocultures are easier to plant, to weed, and to harvest than uneven-aged polycultures, when these operations are carried out mechanically. However, in developing countries, there is often an excess of labour and a shortage of capital and equipment. In such cases, the advantage of machine planting, weeding and harvesting disappears. It is better to employ intensive labour for these operations. Once a decision has been made to employ intensive labour throughout the lifetime of a plantation, the advantage of even-aged monocultures disappears. When people are doing the planting, it is just as easy to plant and cultivate a diverse plantation of various ages than a single-aged monoculture.

For a century or more, elephants have been used to extract teak from the native forests in northern Thailand. Oxen and other beasts of burden are often used in other regions. Where animal power is a traditional means of extraction, and it is not necessary to construct expensive roads at the time of each harvest, an uneven-aged management is much more feasible.

Ecological theory suggests that a diversity of species and of structures may not only be feasible, but also desirable. We suggest that the greatest advances for the future of taungya and plantation forestry could be in trials of mixed species and uneven-aged trees. Where labour is cheap and available, and equipment and capital and land is scarce, diversified taungya may be a partial solution to some of the economic, social and ecological problems which exist in conventional taungya.

## References

Ae, N., Arihara, J., Okada, K., Yoshihara, T. and Johansen, C. (1990) Phosphorus uptake by Pigeon pea and its role in cropping systems of the Indian subcontinent. *Science* 248, 477–480.

Andrews, D.J. and Kassam, A.H. (1976) The importance of multiple cropping in increasing world food supplies. In Papendick, R.I., Sanchez, P.A. and Triplett, G.B. (eds) *Multiple Cropping. American Society of Agronomy Special Publication* 27, pp. 1–10. Cited in Francis (1990).

Batmanian, G.J. (1990) Reforestation of degraded pastures in the Brazilian Amazon: effect of site preparation on phosphorus availability in soil. PhD dissertation, University of Georgia, Athens, USA.

Beer, J. (1987) Advantages, disadvantages and desirable characteristics of shade trees for coffee, cacao and tea. *Agroforestry Systems* 5, 3–13.

Cecilio, P. and Janos, D.P. (1987) *Vesicular-arbuscular mycorrhizas cause maize-solanum polyculture overyielding.* Cited in Vandermeer (1989).

Evans, J. (1982) *Plantation Forestry in the Tropics.* Clarendon Press, Oxford.

Francis, C.A. (1986) *Multiple Cropping Systems.* MacMillan, New York.

Gajaseni, J. and Jordan, C.F. (1990) Diversified agroforestry system. In Kartasubrata, J., Tjitrosomo, S.S. and Umaly, R.C. (eds) *Symposium on Agroforestry Systems and Technologies,* Bogor, Indonesia, 19–21 Sept., 1989. BIOTROP Special Publication No. 39. SEAMEO BIOTROP, Box 17, Bogor, Indonesia, pp. 157–161.

Gliessman, S.R. (1986) Plant interactions in multiple cropping systems. In Francis, C.A. (ed.) *Multiple Cropping Systems.* Macmillan, New York.

Gomez-Pompa, A., Whitmore, T.C. and Hadley, M. (eds) (1991) *Rain Forest Regeneration and Management.* UNESCO and The Parthenon Publishing Group, Carnforth.

Han, X. (1989) The mobilization of aluminum- and iron-bound phosphorus by organic matter decomposition in highly weathered soils. PhD dissertation. University of Georgia, Athens, GA, USA.

Hansen, M.K. (1983) Interactions among natural enemies, herbivores, and yield in monocultures and polycultures of corn, bean, and squash. PhD dissertation, University of Michigan, Ann Arbor, Michigan, USA.

Jordan, C.F. (1987) *Amazonian Rain Forests. Ecosystem Disturbance and Recovery.* Springer Verlag, New York.

Jordan, C.F. (1989) *An Amazon Rain Forest: The structure and function of a nutrient-stressed ecosystem and the impact of slash-and-burn agriculture.* Parthenon Publishing, Carnforth UK and UNESCO, Paris.

Jordan, C.F. and Russell, C.E. (1989) Jari: a pulp plantation in the Brazilian Amazon. *GeoJournal* 19 (4), 429–435.

Lincoln, C. and Isley, D. (1947) Corn as a trap crop for the cotton bollworm. *Journal of Economic Entomology* 40, 437–438.

National Research Council (1989) *Alternative Agriculture.* National Academy Press, Washington DC.

Oldeman, R.A.A. and vanDijk, J. (1991) Diagnosis of the temperment of tropical rain forest trees. In Gomez-Pompa, A., Whitmore, T.C. and Hadley, M. (eds) *Rain Forest Regeneration and Management.* UNESCO and The Parthenon Publishing Group, Carnforth.

Snaydon, R.W. and Harris, P.M. (1979) Interaction belowground – the use of nutrients and water. In Willy, R.W. (ed.) *Proceedings of the International Workshop on Intercropping.* Hyderabad, India. ICRISAT, Andhra Pradesh, India.

Trenbath, B.R. (1979) Light-use efficiency of crops and the potential for improvement through intercropping. In Willy R.W. (ed.) *Proceedings of the International Workshop on Intercropping.* Hyderabad, India. ICRISAT, Andhra Pradesh, India.

Vandermeer, J.H. (1989) *The Ecology of Intercropping.* Cambridge University Press, Cambridge.

Vandermeer, J.H. (1990) Intercropping. In Carroll, C.R. Vandermeer, J.H. and Rosset, P. (eds) *Agroecology.* McGraw-Hill, New York.

Whittaker, R.H. (1975) *Communities and Ecosystems.* MacMillan, New York.

# PART II
## Regional Perspectives on Taungya

# Taungya in Thailand: Editors' Note

The taungya system of reforestation as originally practised in Burma (Chapter 2), was adopted in 1906 by the Royal Forestry Department of Thailand in the province of Prae in Northern Thailand. The attempt was not successful because of conflicts between the interests of the peasants in crop production and the foresters who were concerned with reforestation and tree production. As a result, the system was abandoned after a few years, and reforestation was carried out by hired labour.

After the Second World War, the population of Thailand increased rapidly, and there was increasing need for agricultural land. As a result, much land was deforested by slash and burn cultivators. Between 1961 and 1988, forest cover diminished from 53% to 28% of total land area (see the Table on p. 86).

In addition to loss of forest land due to conversion to agriculture, many of the best stands of forest trees, especially teak, were over-harvested or illegally cut, resulting in diminished stocks and inadequate reproduction. Because teak is an important export commodity for Thailand, loss of teak forests results in decreases in revenue for the country.

To remedy the problem of inadequate supplies of teak and the need for employment of landless peasants, a modified version of taungya called the Forest Village System was introduced in Thailand. One major initiative was begun in 1967 sponsored by the Forest Industry Organization of Thailand. The primary objective was production of trees. A second effort was begun in 1976 by the Royal Forestry Department. In these villages, more emphasis was given to social and environmental concerns.

In the following two chapters, the two approaches to the Forest Village System of taungya are described by professionals who have worked with the two respective agencies.

Natural forest areas in various regions of Thailand.

| Region | Total land area | Natural forest area (ha) | | | | | | |
|---|---|---|---|---|---|---|---|---|
| | | 1961 | 1973 | 1976 | 1978 | 1982 | 1985 | 1988 |
| North | 16964429 | 11627500 (68.54) | 11359500 (66.96) | 10232700 (60.32) | 9493700 (55.96) | 8775600 (51.73) | 8412600 (49.59) | 8040227 (47.39) |
| East | 3650250 | 2116300 (57.98) | 1503600 (41.19) | 1263100 (34.60) | 1103700 (30.24) | 800000 (21.92) | 799000 (21.89) | 783382 (21.46) |
| Northeast | 16885433 | 7090400 (41.99) | 5067100 (30.01) | 4149400 (24.57) | 3122100 (18.49) | 2588.600 (15.33) | 2422400 (14.35) | 2369335 (14.03) |
| Central | 6739870 | 3566050 (52.91) | 2397000 (35.56) | 2182600 (32.38) | 2042600 (30.31) | 1851600 (24.47) | 1722800 (25.56) | 1724435 (25.59) |
| South | 7071519 | 2962600 (41.89) | 1843500 (26.07) | 2013900 (28.48) | 1760300 (24.89) | 1644200 (23.25) | 1548500 (21.90) | 1462963 (20.69) |
| Whole country | 51311501 | 27362850 (53.33%) | 22170700 (43.21%) | 19841700 (38.67%) | 17522400 (34.15%) | 15660000 (30.52%) | 14905300 (29.05%) | 14380349 (28.03%) |

Source: Royal Forestry Dept. Bangkok, Thailand.
Note: All data were derived from LANDSAT except for the 1961 data which came from aerial survey.
Values in parenthesis are percentages of the total land area in each region.

Chapter 8

# Taungya in Thailand: Perspective of the Forest Industry Organization

Montee Phothitai

*Planning and Statistics Sub-division, Forest Industry Organization, Bangkok, Thailand*

The Forest Industry Organization (FIO) was set up in 1947 to serve the needs of Thailand in forest industry, and to promote commercialization of timber harvest, reforestation, and forest research. In 1967, the FIO initiated a project called the Forest Village System, to try to solve the problems of land degradation, lack of land for peasants, and the need for reforestation of valuable timber species. The objective of the Forest Village System is to settle landless people, and to assure a source of labour for plantation work.

The project was based upon the 'taungya' approach that had been tried briefly in Thailand in the early part of this century. In that approach, peasants were allowed to cultivate subsistence crops for the first 2 years of a forest plantation. The system was abandoned, in part because of conflicts between the needs of agriculture and forestry.

The modern Forest Village System under FIO differs from the old taungya system in the wage benefits, land utilization rights, and socioeconomic improvement schemes, e.g. permanent settlement, water, electricity, education and health care. Each family is allotted 1.6 ha of land each year. The land is to be cleared and planted with forest trees, and cultivation of crops is permitted between the tree seedlings. In addition to the 1.6 ha which is allotted each year, an additional 0.16 hectares of land in the village is given for a dwelling and backyard gardening.

Participants in the system are given financial rewards. A quota of two labourers from each family is recruited to work in plantation activities. This employment adds substantially to the income of the family. For every 0.16 ha that are planted, the family is given a reward of US$8.00. Also, if any family stays with the Forest Village System for three consecutive years and plants over 4.8 ha, they are paid a bonus of US$60. After the third year, for every additional year they remain and continue to plant their 1.6 ha per year, they receive an extra US$48. Total income per family per year for the various regions of Thailand is given in Table 8.1. Daily wages constitute a very important portion of total income.

With this plan, plantations are established at the rate of 160 ha per year in each village of 100 families. The area allotted to a village should be large enough so that a whole rotation can be completed before there is a need for additional land for new plantations. Initially, rotations for teak in the north of Thailand were planned to be for 60 years, but they may be scaled back to 30 years, depending on site quality and economic considerations.

The Forest Industry Organization annually allocates over US$5 million to establish forest plantations in Thailand. At present, there are 42 forest villages in the plantation programme, 20 of which are in the north, 13 in the east and northeast, and 9 in the western and southern parts of the country (Table 8.2). These plantations serve the Organization's aim to bring benefits of job-opportunity programmes and community development to inhabitants in poor, rural communities.

## Types of Taungya

The taungya systems of the FIO in Thailand can be grouped into three types as follows.

### Forest trees and agricultural crops

In the north, east, northeast, and west, peasants in the FIO Forest Village System usually plant agricultural crops in the plantations to earn supplementary income. Preferred crops have a high productivity, and are in demand economically. In order to make the plantation better adapted to agricultural crops, the original tree spacing, 4 × 4 m is being modified to 2 × 8 m.

In the north, the forest tree is teak (*Tectona grandis*). The main crops are maize and upland rice. Minor crops are groundnut, pumpkin, chilli, red bean, mung bean and cotton.

In the northeast and east, the forest trees are *Eucalyptus camaldulensis* and *Melia azedarach*. The main crop is cassava, and minor crops are maize, kenaf, upland rice, groundnut, sorghum, bean, mung bean, red bean, and chilli.

In the west, the forest trees are teak, *Eucalyptus camaldulensis* and *Melia azedarach*. The main agricultural crop is maize, and minor crops are upland rice, cotton, and castor bean. A more complete list of combinations for each region is given in Table 8.3.

**Table 8.1.** Income (US$) of the forest villagers of FIO (1988).

| Region | No. of families | Sources of income | | | | Total | Total income/ family/year |
|---|---|---|---|---|---|---|---|
| | | Daily wages | Taungya compensation | Taungya bonuses | Agro-crop income | | |
| North | 899 | 415 567 | 15 336 | 4615 | 195 788 | 631 306 | 702.23 |
| East and northeast | 743 | 206 787 | 3701 | 74 884 | 129 052 | 414 424 | 557.77 |
| West and south | 302 | 237 567 | 6597 | 22 329 | 73 077 | 339 570 | 1124.40 |
| Total | 1944 | 859 921 (62.08%) | 25 634 (1.85%) | 101 828 (7.35%) | 397 917 (28.72%) | 1 385 300 | |

Source: Forest Industry Organization, 1988.

**Table 8.2.** Forest village, location and areas established by FIO (1988).

| Region | Forest village name | Location (district, province) | Year established | No. of families | Population | Area established (ha) |
|---|---|---|---|---|---|---|
| North | 1. Tung Kawien | Hangchat, Lampang | 1968 | 37 | 169 | 2443.52 |
| | 2. Maemai | Muang, Lampang | 1971 | 50 | 209 | 3283.52 |
| | 3. Maemoh | Maemoh, Lampang | 1968 | 101 | 425 | 3038.72 |
| | 4. Mae Jang | Mae ta, Lampang | 1968 | 67 | 303 | 2888.80 |
| | 5. Khun mae Kammee | Rong Kwang, Prae | 1970 | 39 | 231 | 3267.84 |
| | 6. Khao Krayang | Wangthong, Pitsanuloke | 1968 | 30 | 149 | 2560.32 |
| | 7. Mae Sai Kham | Muang, Lampang | 1971 | 78 | 374 | 2373.12 |
| | 8. Mae Hor Pra | Mae taeng, Chiengmai | 1972 | 44 | 190 | 1479.36 |
| | 9. Mae Li | Li, Lumpun | 1971 | 14 | 72 | 2376.80 |
| | 10. Ban dan lan hoy | Bandan, Sukhothai | 1971 | 43 | 255 | 2320.00 |
| | 11. Mae Jaem | Mae Jaem, Chiengmai | 1973 | 95 | 401 | 1164.48 |
| | 12. Lad yao | Ladyao, Nakornsawan | 1977 | 55 | 247 | 913.28 |
| | 13. Mae Lamao | Mae sod,Tak | 1974 | 9 | 40 | 1132.48 |
| | 14. Srisatchanalai | Srisatchanalai, Suknothai | 1975 | 60 | 226 | 2103.36 |
| | 15. Maeyom-Maepaeng | Wangchin, Prae | 1978 | 7 | 32 | 298.40 |
| | 16. Wangchin | Wangchin, Prae | 1977 | 63 | 288 | 1559.88 |
| | 17. Mae Saroy | Wangchin, Prae | 1978 | 32 | 156 | 1717.44 |
| | 18. Khao Kana | Chon daen, Petchabun | 1978 | 12 | 52 | 720.32 |
| | 19. Luang sankampaeng | Sankampaeng, Chiengmai | 1979 | 50 | 197 | 744.00 |
| | 20. Mae Kampong-Mae Sai | Rongkwang, Prae | 1979 | 13 | 73 | 655.36 |

| | | | | | | |
|---|---|---|---|---|---|---|
| East and northeast | 21. Somdet | Nuang, Kalasin | 1975 | 232 | 1146 | 2001.28 |
| | 22. Sakaeo | Wangnamyen, Prachinburi | 1974 | 46 | 248 | 898.40 |
| | 23. Wathananakorn | Wathananakorn, Prachinburi | 1978 | 16 | 80 | 427.04 |
| | 24. Sopisai | Sopisai, Nongkai | 1977 | 15 | 93 | 181.60 |
| | 25. Khunharn | Khunharn, Srisaket | 1974 | 77 | 330 | 746.24 |
| | 26. Tepsathit | Tepsathit, Chaiyaphum | 1978 | 58 | 225 | 742.72 |
| | 27. Pibulmungsaharn | Pibulmungsaharn, Ubolrachathani | 1976 | 84 | 470 | 1376.32 |
| | 28. Namsom | Namsom, Udornthani | 1976 | 13 | 79 | 318.24 |
| | 29. Khonsarn | Khonsarn, Chaiyaphum | 1979 | 90 | 424 | 704.16 |
| | 30. Kasetsomboon | Kasetsomboon, Chaiyaphum | 1979 | 27 | 101 | 574.08 |
| | 31. Dankhuntod | Dankhuntod, Nakornrachasima | 1980 | 21 | 102 | 911.84 |
| | 32. Manjakiri | Manjakiri, Khonkaen | 1980 | 44 | 210 | 1354.88 |
| | 33. Sangkha | Sangkha, Surin | 1980 | 20 | 102 | 244.32 |
| West and south | 34. Salui-Tasae | Tasae, Chumporn | 1978 | 29 | 148 | 893.76 |
| | 35. Sinpun | Prasaeng, Surathani | 1978 | 19 | 89 | 327.20 |
| | 36. Klongtom | Klongtom, Krabi | 1975 | 17 | 106 | 2121.12 |
| | 37. Karnchanadit | Donsak, Surathani | 1974 | 34 | 135 | 898.88 |
| | 38. Prasaeng | Prasaeng, Surathani | 1975 | 70 | 344 | 1279.36 |
| | 39. Wangviset | Wangviset, Trang | 1975 | 9 | 77 | 1483.84 |
| | 40. Tongpaphum | Tongpaphum, Kanchanaburi | 1979 | 70 | 327 | 1935.52 |
| | 41. Ongpra | Danchang, Supanburi | 1979 | 46 | 232 | 937.92 |
| | 42. Ongkot | Danchang, Supanburi | 1984 | 8 | 53 | 340.96 |
| | Total | | | 1944 | 9210 | 57740.48 |

Source: Forest Industry Organization, 1988.

**Table 8.3.** Tree–crop combinations in the Modified Taungya System of the Forest Village of FIO (1988).

| Region | Forest village name | Tree species planted | Intercropped species planted |
|---|---|---|---|
| North | 1. Mae Hor Pra | Te g | Or s, Ar h, Go s, Gl m |
| | 2. Mae Li | Te g | Ze m, Gl m, Or s, Ar h, Al a |
| | 3. Mae Jaem | Te g | Gl m, Ze m, Ve c |
| | 4. Lad yao | Te g | Ze m, Ma e, Vi r |
| | 5. Srisatchanalai | Te g | Gl m, Ze m, Oc b |
| | 6. Wangchin | Te g | Ze m, Vi r, Be h, Or s Cu m, Go s, Ca a, Se o |
| | 7. Luang sankampaeng | Te g | Ze m, Ni t, Sa o, Mu s, Ar h |
| East and northeast | 8. Somdet | Eu c | Ma e, Or s |
| | 9. Sakaeo | Eu c | Ma e, Or s, Co s, Ze m, Vi r So b |
| | 10. Wathananakorn | Eu c | Ma e, Ze m, Vi r |
| | 11. Khunharn | Eu c | Ma e, Co s, Ar h, Cu m |
| | 12. Tepsathit | Eu c | Ma e, Co s, Ze m |
| | 13. Namsom | Eu c | Ma e, Mu s |
| | 14. Khonsarn | Eu c | Ma e, Ze m, Gl m, Vi u, Ca a |
| | 15. Manjakiri | Eu c | Ma e |
| West and south | 16. Salui-Tàsae | Di a, Le l | Co r, He b |
| | 17. Sinpun | Le l | Or s, He b |
| | 18. Klongtom | Ca e, Eu c, Le l | He b, An o |
| | 19. Karnchanadit | Pa j | Co r, He b |
| | 20. Prasaeng | Le l | Co r, He b |
| | 21. Wangviset | Le l | He b |
| | 22. Tongpaphum | Te g, Le l | Ze m, Gl m, Vi r, Co s, He b |
| | 23. Ongpra | Eu c | Ze m |

Source: Forest Industry Organization, 1988.

**Agricultural crop**
Maize: Ze m, *Zea mays*
Upland rice: Or s, *Oryza sativa*
Pumpkin: Cu m, *Cucurbita moschata*
Groundnut: Ar h, *Arachis hypogaea*
Sesame: Se o, *Sesame orientale*
Mungbean: Vi r, *Vigna radiata*
Cassava: Ma e, *Manihot esculenta*
Cotton: Go s, *Gossypium* spp.
Soy bean: Gl m, *Glycine max*
Chilli: Ca a, *Capsicum annuum*
Tobacco: Ni t, *Nicotina tabacum*
Kenaf: Co s, *Corchorus* spp.
Wax gourd: Be h, *Benincasa hispida*
Shallot: Al a, *Allium ascalonicum*
Sweet basil: Oc b, *Ocimum basilicum*
Sorghum: So b, *Sorghum bicolor*
Vegetable crops: Ve c

**Tree species:**
Teak: Te g, *Tectona grandis*
Yang: Di a, *Dipterocarpus alatus*
Ipil-ipil: Le l, *Leucaena leucocephala*
Eucalypt: Eu c, *Eucalyptus camaldulensis*
Beefwood: Ca e, *Casuarina equisetifolia*
Pa j, *Parkia javanica*

**Horticultural/fruit crop:**
Banana: Mu s, *Musa sapientum*
Sugarcane: Sa o, *Saccharum officinarum*
Coffee: Co r, *Coffea robusta*
Cashew: An o, *Anacardium occidentale*
Pararubber: He b, *Hevea brasiliensis*

## Forest trees and horticultural crops

In Southern Thailand, the people have a tradition of working with horticultural crops like coffee, rubber, cashewnut, and other fruit trees. For this reason, FI0 reforestation projects in the Southern region maintain forest trees combined with horticultural crops instead of with agricultural crops. The forest trees planted are *Dipterocarpus alatus*, *Leucaena leucocephala*, and *Parkia javanica*. Horticultural crops are rubber, coffee, and cashewnut.

In general, tree crops and fruit trees grow better than annual grains in the Southern region, and provide a better income. Therefore, they are preferred by the local peasants. To accommodate fruit trees between the forest trees, the forest trees are planted farther apart than in the north.

## Animals and trees

The practice of combining animals with forestry has been introduced at a number of FIO plantations as an additional source of income for participating families. Two types of animal/forest combinations are practised.

**1.** Cattle raising or 'silvopastoral agroforestry'. This involves controlled grazing of forest vegetation and understory weeds in plantations. It benefits the trees by eliminating, to some degree, forest fire hazards. Cattle raising also provides substantial additional income. However, stocking rate must be carefully controlled, and consideration must be given to forest tree species, their size, age, and spacing, as well as adequate water for the cattle.
**2.** Beekeeping or 'apisilviculture'. Native bees are raised in hives, and gather nectar from the flowers of forest trees. Where beekeeping is part of the Forest Village Programme, careful attention must be given to selecting trees that provide nectar, and that there is adequate water near the site.

At present, FIO plantations in the southern and western regions maintain beekeeping as extension programmes for forest village families to improve their standard of living.

The combinations of trees and crops in the Forest Village System differ from village to village, and from region to region due to differences in climate, soil fertility, market demand, and skills and preferences of the cultivators. The taungya agricultural systems change and are improved depending on the market demand and productivity. Rubber with pineapple is becoming more popular in the south, because cultivators are more eager to grow cash-crops than food crops.

Crops planted also depend on seasonality. Maize is usually cultivated in the rainy season, from May to August. When the maize is harvested in the west, sorghum is planted because it is more drought tolerant. Occasionally, maize and sorghum are interplanted on the same land in the north.

Vegetables like beans, pumpkins and sesame are not normally interplanted with tall cereals such as maize or sorghum, because there would be too much competition for light. However, where grains are sparsely planted, such interplanting does occur.

Chapter 9

# Taungya in Thailand: Perspective of the Royal Forestry Department

Ob Preeyagrysorn*

*Payap University, Chiangmai, Thailand*

The Taungya System can be considered as a step in the process of transformation from shifting cultivation to agroforestry. Shifting cultivation is a sequential system of growing woody species and agricultural crops, whereas taungya consists of simultaneous combination of the two components during the early stages of forest plantation establishment. Although wood production is the ultimate objective in the taungya system, the immediate motivation for its practise is food production for the local people.

From the point of view of soil management, both taungya and shifting cultivation systems are similar. Agricultural crops are planted to take advantage of the improved soil fertility built up by the woody plant components in shifting cultivation. The length of the agricultural cycle can last only as long as the soil sustains a reasonable yield of crop. In shifting cultivation, crop production is often limited by decline in soil fertility. In Taungya, crop production is limited by the space between trees that is free from competition. This space decreases as the trees grow.

Taungya, as practiced in Burma and Thailand in the early part of this century, has been criticized as a labour-exploiting system. It capitalized on the need of the poor forest farmer for food. It took advantage of his willingness to offer labour to plant trees for no pay, in return for the right to raise much needed food crops for only a short period of time. This was one of the reasons that the system was abandoned.

*Formerly with the Royal Forestry Department of Thailand, Bangkok.

## Forest Village Programme

Since the 1950s, problems in rural areas of Thailand have increased. As a result of the heavy clearing of vast areas of national forest reserves by squatters for agricultural purposes, droughts and flash floods became a common phenomena. There was a shortage of timber, supplies were not adequate for domestic use, and exports declined resulting in a decline in national revenues. These factors affected the socioeconomic and political conditions in Thailand. Therefore, on 29 April 1975, the cabinet of Ministers directed the Royal Forestry Department (RFD) to solve the problems through the allocation of land in degraded national forests reserves to taungya in the Forest Village Programme.

The modern programme differs from taungya in Burma and Thailand near the turn of the century in several ways. In the traditional taungya system, the farmers were allowed to grow crops for 1 or 2 years among the rows of trees. However, the farmers were not supplied with any facilities, nor given any economic or social assistance. As a result, the peasants were discontent with the system, and would not give up their nomadic existence. The system was often faced with shortages of labour for plantations.

In the modified modern system of taungya, a rural development scheme is integrated into the reforestation scheme. In this way, the labour force is stabilized, and a continuous labour force is ensured for the work of plantation.

While the Forest Village System organized by the Royal Forestry Department (RFD) is similar in some ways to the system organized by the Forest Industry Organization (FIO), there is a difference in the primary objective of the two organizations.

**1.** FIO. The primary goal of the Forest Village System organized by the FIO is to ensure that there is a sustained yield of timber. Resettlement of landless farmers from land belonging to the government is another important reason. Improvement of the socioeconomic situation of local farmers is a secondary objective. Administration of the Villages is only by the FIO.

**2.** RFD. In contrast to the FIO villages where timber production is the main goal, raising the standard of living of rural people, and improving their socioeconomic status is the primary objective in the RFD villages. Other objectives of the RFD programme relate to environmental and social concerns. They include: increasing public security; rehabilitating degraded watersheds; resettling landless farmers on a permanent basis; stopping encroachment on national forest reserves; rehabilitating degraded forest reserves and increasing the timber production in these areas; assisting the farmers with agricultural extension, marketing services, etc.

In the modern villages, in which there are about 100 families, a number of local governance and development organizations have been established.

These institutions assist village officials in managing community affairs, encourage participation in community-wide development activities, and promote the government's social, economic and internal security policies. The model used in designing the Forest Village Programme has provided for a government-approved village headman, a village committee, and a development fund managed by elected neighbourhood members, a rice bank, young adult club, agricultural credit programme, and paramilitary organizations created under the national programme of Political Groups for National Security. In addition, there are schools, medical clinics, temples, and other social services. Water and electricity also are supplied.

## Programme Specifications

The resolution that specified the establishment of forest villages in degraded forest reserves included the following basic guidelines.

**1.** Resettlement of squatters will be in a group with an elected leader and a committee for village self-government.
**2.** The government will provide land for farming at the rate of 15 rai per family (6.25 rai = 1 hectare). Of that land, 1 rai will be in the village and can be used to build a house and establish a home garden. The other 14 rai will be in an area in which crops can be raised along with tree plantations. The amount of land will be more or less depending on the number of landless people. A user permit will be given for this land. Land is located in non-watershed areas. The property is inheritable but non-negotiable. This is to prevent acquisition of large tracts by landlords. Membership in the villages is voluntary.
**3.** In each village, the Royal Forestry Department and other government agencies will provide suitable vocational training, irrigation, feeder roads, schools, health clinics, agricultural credits, marketing services, and housing areas.
**4.** Priority will be given to village members for employment opportunities in the reforestation projects around the villages.
**5.** After the village is established, it will be organized as an agricultural cooperative under the aegis of the Cooperative Promotion Department, with the same rights and privileges as other cooperatives. The Royal Forestry Department will issue a long-term lease of land to the village cooperative as required.

## Operating Procedures

A land holding survey will be carried out to assess the number of squatters and the nature of the holdings in the area for reforestation. This information is necessary to determine the right of the squatters to obtain land in the village. The procedure at this stage involves squatter registration, on-site examination of land holdings, verification of history of squatters on the site, and determination of rights of this person and family to move into the village.

Villages are organized and built according to a model plan. The plan includes demarcation of living and farming plots, providing for a source of water, and construction of public utilities. It also includes provision for assisting with emigration into the village, site development, and maintenance.

The Public Health Department will establish a new health centre or expand the existing centre. The Provincial Administration Office will either set up a new school or expand the existing school. Financial assistance will be provided through the cooperation of the Bank for Agriculture and Cooperatives or other financial institutions for both housing and farming loans.

The Department of Agricultural Extension will promote appropriate agricultural practices. Seedlings will be distributed to village members for planting in their areas. Village members will be hired for nursery work. In watersheds and areas unsuitable for agriculture, reforestation for rehabilitation will be carried out using labour from the village. Community forestry for supplying the village with firewood, fence posts, piling posts, etc., will be established. Nursery work, reforestation and community forestry will be the major sources of employment for village members.

## Administrative Committee

The administration of the forest villages comes under two levels of committee, the Local and the National committees. The functions of these committees are:

**1.** to supervise the allocation of land in the Forest Village Programme, as specified by target and by plan;
**2.** to solve problems encountered in the Programme;
**3.** to maintain law and order in the Programme area;
**4.** to provide security, education, vocational training, and other public functions;
**5.** to publicize the Programme.

Policy questions which cannot be answered by the Local Committee are referred to the National Committee for final decision.

## Present Programme

At present, 170 Forest Villages have been established throughout Thailand under the RFD programme. Of these, 50 villages are in the northern part of the country, 79 in the east and northeast, and 41 in the west and southern regions.

### EDITORS' NOTE

Continued forest land encroachment and problems in implementing the Forest Village Programme have prompted renewed consideration of alternative solutions to these problems. In 1981, the Thai cabinet granted formal approval of the National Forest Land Allotment (STK) Project as a remedial solution. This project was primarily motivated by the slow progress in developing new forest villages, the high costs of establishing these villages, and inadequate budgets for this work.

The STK Project has had objectives similar to those of the Forest Village Program: to grant agricultural land entitlements in reserved forests, to promote reforestation and conservation of degraded forest land, to improve relations between forestry officials and local residents, and to discourage repeated migration and encroachment on reserved forest land. STK centers were established in Royal Forestry Department regional offices to plan, monitor, and implement activities in specific project locations.

The STK Projects differ from the Forest Village Programme in a number of important respects. No land was identified or allocated for houselots, village paths and roads were not developed, and no village infrastructure or services were provided. Land allocations were limited to a maximum of 2.4 ha per household, but previously claimed and cleared land could continue to be cultivated by paying an annual fee. This land was not confiscated for reallocation, and this allowed many of the original claimants to retain the right to use land in which they had already invested labour and capital.

Hafner and Apichatvullop (1990) carried out a study to compare the Forest Village Programme at Phu Hang with the STK Project at Non Amnuay. They state that the Forest Village Programme has been plagued by conflicts between villages, foresters responsible for managing the programme, and district officials. These disputes illustrate a variety of problems with poor communication about the programme's goals and objectives, implementation policies, and the responsibilities of different government agencies.

In contrast, say Hafner and Apichatvullop, the STK Project in Non

Amnuay has been marked by much less social and institutional conflict. Disputes over original land claims were mainly resolved before land entitlements were granted. The entitlement process has also enabled villagers to retain land originally claimed and developed, to cultivate more than 2.4 ha, and has given landless households the opportunities to rent land.

This sort of contrast may not occur in all the forest villages, nor STK projects. However, we felt it important to point out that in at least some cases, problems remain, but efforts are continuing to resolve the problems.

## Reference

Hafner, J.A. and Apichatvullop, Y. (1990) Migrant farmers and the shrinking forests of Northeast Thailand. In Poffenberger, M. (ed.) *Keepers of the Forest: Land Management Alternatives in Southeast Asia.* Kumarian Press. W. Hartford, Connecticut, pp. 187–219.

Chapter 10

# Taungya Cultivation in Java, Indonesia: Agrisilvicultural and Socioeconomic Aspects

Hasanu Simon[1] and K.F. Wiersum[2]

*[1]Forestry Faculty, Gadjah Mada University, Yogyakarta, Indonesia. [2]Department of Forestry, Wageningen Agricultural University, POB 342, 6700 AH Wageningen, The Netherlands*

The Indonesian island of Java covering an area of 13.2 million hectares is one of the most populated regions in the world. In 1930 its population was 35 million, but this has now increased to about 100 million. This means an average population density of around 750 persons $km^{-2}$. Owing to this high population density, the average size of farmholdings is very small: on average 0.48 ha consisting of 0.26 ha dryland fields and home-gardens and 0.22 ha of irrigated rice fields. Forests cover 23% of the land area. Several million people live in villages surrounding these forest areas. Most of them are small farmers or farm labourers; about half of them are under-employed or unemployed. Obviously, such social conditions result in a great pressure on forest lands. Therefore it has been stipulated in the official forest policy, that the forest areas should not only be managed for the production of timber and commercial non-timber forest products, but that forest management should also be directed at increasing employment opportunities and rural community development (Atmosoedardjo and Wahyudi, 1980; Wirjodarmodjo and Bratamihardja, 1984).

Taungya is being used as one of the most important tools for implementing this policy. Indeed, taungya has been practised in Java for over a century. The system was first introduced in Central Java in 1873. Already in 1881 it was included in the official forestry regulations as an obligatory technique for establishing teak (*Tectona grandis*) plantations. Since this period the techniques used have gradually evolved in response to changes in socioeconomic conditions. Originally, taungya was applied to establish teak plantations on lands previously occupied by natural forests. These areas were mostly situated in hill regions consisting of tertiary sedimentary deposits, where soils had a relatively modest fertility. The population density in these areas was relatively low. Increasingly, the technique became used also for the establishment of second generation plantations and afforestation of degraded areas. Since the middle of the 20th century the technique has also been extended to fertile volcanic uplands for establishment of pine and agathis plantations. In these areas there is often intensive competition for land between commercial horticulture and forest plantations (Wiersum, 1982). In addition, the system has also been modified for use in mangrove forest areas, where forest regeneration is combined with fish cultivation (Wiersum, 1981).

The population density in the regions with teak forests has also gradually increased and several efforts have been made to adjust the taungya system to current socioeconomic conditions. This chapter reviews the evolution of the taungya system in teak plantation areas in Java. They occupy 32% of the total forest area on Java and form the major part of the commercial timber estate. First, a description of the main characteristics of the traditional taungya system, as practised up to the 1970s, will be given. This will be followed by a discussion on the characteristics of intensified taungya cultivation, which started around 1975. Attention will also be given to further adjustments which are now being tried out. Finally, an assessment will be made of the place of taungya in respect to the overall forest management system.

## Traditional Taungya in Java

Taungya was introduced at the end of the 19th century for several reasons (Becking, 1928; Beekman, 1948):

**1.** to decrease the establishment costs of teak stands;
**2.** to obtain an additional income from agriculture during the juvenile stage of tree plantations;
**3.** to achieve better maintenance of the young tree stands;
**4.** to reclaim wasted lands by means of agriculture before the establishment of tree plantations;
**5.** to contribute to solving local shortages of good agricultural lands.

Obviously, taungya was mainly considered as an effective means of reforestation. Its contribution to improving the welfare of local people was subsidiary to the successful establishment of timber plantations. This view prevailed up to the middle of the 20th century; e.g. in 1953 it was stated that ‘a prerequisite of taungya is forestry, (. . .) the own objectives of the forest enterprise may not be hindered by increased food production’ (Hellinga, 1953).

The major characteristics of traditional taungya as evolved in the middle of the 20th century were as follows (Kartasubrata, 1979). First a contract was made between individual farmers and the Forest Service, in which the area to be planted, period of agricultural cultivation and allowed crops were specified, as well as the contract sum to be paid by the Forest Service for tree planting and tending. Originally a farmer was free to choose the area he wanted to cultivate, but in the 1950s a 0.5 ha plot was usually allocated per family. After clearing, these plots were cultivated with crops such as dryland rice, maize, pepper, peanut and soybean. The growing of tuber crops such as cassava and potato was originally forbidden, as tree growth was negatively affected by their cultivation (Table 10.1); gradually this rule has been relaxed, however. The food crops were interplanted between alternating rows of direct-seeded teak (spaced 3 × 1 m) and *Leucaena leucocephala.* The last species was used as a permanent cover-crop which was regularly pruned. (After the recent infestation of the leucaena psyllid now also other cover crops such as *Tephrosia* spp. are used.)

Taungya cultivation was allowed on slopes up to 30% with the provision that on sloping lands light erosion control structures in the form of trash or stone ridges should be established. Agricultural cultivation could continue for 2 years. Afterwards the taungya farmers continued to tend the trees for another 3 years. The tree-tending activities consisted of selection of leaders in case more than one seed per planting hole (of the 2–3 seeds used) germinated, infilling in case of insufficient germination and early mortality, regular coppicing of *Leucaeana* (twice a year) and weeding. Supervision over all the work was carried out by a foreman of the Forest Service assisted by four volunteers selected from the contracted farmers; together they supervised an area of 10–12 ha (Hardjosoediro, 1979).

The use of the taungya system was very profitable for the Forest Service as it decreased labour costs for tree planting from 86–116 man-days $ha^{-1}$ for planting techniques involving cleaning lines and making planting holes to only 62 man-days $ha^{-1}$. The other 58 man-days $ha^{-1}$

**Table 10.1.** Effect of different food crops on 14-month-old teak growth.

| Crop | Height (% of control) | Girth (cm) |
|---|---|---|
| Control | 100 | 11.9 |
| Rice and peanuts | 95 | 10.3 |
| Peanuts/pepper | 94 | 11.6 |
| Maize | 92 | 10.8 |
| Rice | 96 | 10.0 |
| Cassava | 74 | 10.0 |

Source: Coster and Hardjowasono, 1935.

needed for taungya cultivation were contributed free-of-charge by the taungya farmer in exchange for his rights to agricultural produce. Originally, these taungya farmers were also part-time casual employees of the Forest Service, carrying out logging and maintenance activities during the agricultural off-season. But as more farmers became involved in taungya cultivation, gradually such combined employment became less common.

For several decades this organization of taungya has given satisfactory results, providing food and labour opportunities to the local population as well as benefits to the Forest Service. However, in the 1950s and 1960s the system became gradually less able to cope with increasing pressures of population growth on the forest lands. The combined effects of increased demands for both timber, fuelwood and fodder products for local people, and for increased food production and additional labour opportunities of farmers whose landholdings steadily decreased in size, caused a gradual decline in the quality of teak plantations (Fig. 10.1). First officials tried to adjust to such pressures by allocating smaller taungya plots to more peasants; in some cases plots became as small as 0.1–0.25 ha. But this practice resulted in overcropping of these small plots and neglect of the trees, while control became less effective because more contractors had to be supervised. Also, increasingly, uncontrolled livestock grazing took place after the end of the agricultural cropping period (Hardjosoediro, 1979).

As a result of these developments young plantations failed more often, and required replanting. Also, older plantations became degraded by illegal cutting and overgrazing and had to be rejuvenated. Thus, in 1975 it was estimated that 300 000 ha (or 30%) of teak forest area had become unproductive (Simon, 1989). Consequently, a change took place in respect to the kind of lands on which taungya could be practised. Originally these were newly reclaimed lands with a moderate fertility. But increasingly more or less degraded and impoverished forest lands were included in the taungya schemes. As a result, agricultural yields declined. In the 1930s crop yields in taungya cultivation were reported to be higher than under the average Javanese farming conditions (Coster and Hardjowasono, 1935), but around 1980 the opposite was the case (Table 10.2). Owing to these low yields, the pressure on the forest lands once again increased and in several places a vicious circle of low agricultural yields, neglect for tree crops resulting in failed plantations, land degradation, and repeated taungya cultivation evolved. In the 1970s it was recognized that in order to break through this cycle of decreased agricultural yields and failed plantations it was necessary to adjust the taungya cultivation system by intensifying the agronomic practices.

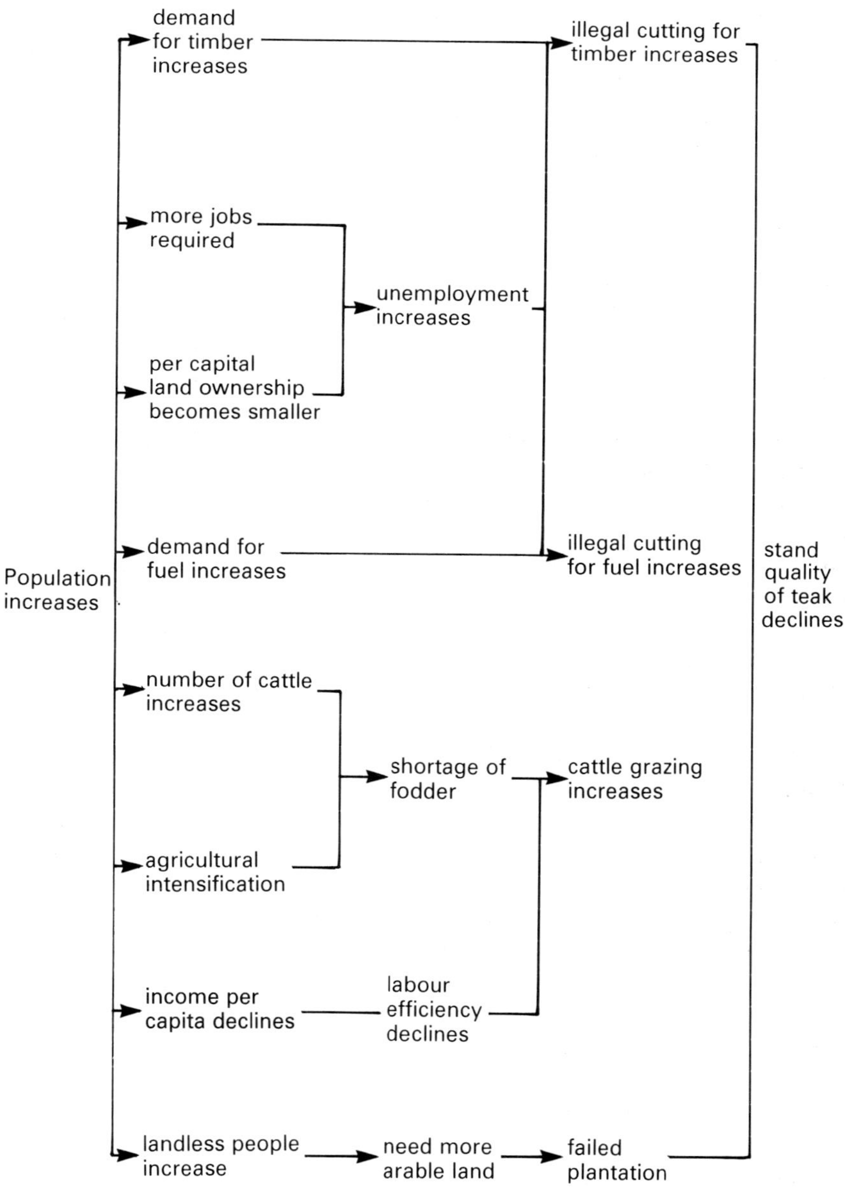


**Fig. 10.1.** Effect of population growth on decline of the quality of teak stands on Java (Simon, 1989).

**Table 10.2.** Selected agricultural yields (kg ha$^{-1}$) under taungya cultivation (a) and average Javanese farming conditions (b).

| Crops | 1935 | | 1980 | | Intensified taungya |
|---|---|---|---|---|---|
| | a | b | a | b | |
| Dryland rice | 2160 | 1200 | 1250–1690 | 1800 | 3250 |
| Maize | 2000 | 990 | 1250–1500 | 1640 | 2250 |

Sources: Coster and Hardjowasono, 1935; Sabarnudin, 1988.

## Intensified Taungya Cultivation

In the 1960s, in Indonesian agriculture, increasing use was made of high-yielding crop varieties and modern cropping practices. In 1972 the State Forest Enterprise decided to test the feasibility of such agricultural improvements for taungya cultivation. After a first trial had proved successful with dryland rice production increasing from 700 to 3740 kg ha$^{-1}$, it was decided to develop this system further. On the basis of research carried out by the Forestry Faculty of the Gadjah Mada University (Wiersum, 1981) a system for intensified taungya cultivation was developed with the following characteristics (Kartasubrata, 1979; Wiersum, 1982):

1. use of selected high-yielding varieties of the main food crops;
2. better practices for land preparation and soil tillage;
3. use of fertilizers;
4. control of pests and diseases;
5. correct timing of planting and fertilizing with respect to the rainfall.

This system provided good results and it was gradually applied over increasing areas. With the use of selected superior crop varieties, fertilization in the range of 90–100 kg urea and 60–150 kg TSP per hectare, if necessary together with the use of insecticides, yields of dryland rice improved from about 700–1500 kg ha$^{-1}$ to 2000–3000 kg ha$^{-1}$ and of maize from 1000–1500 kg ha$^{-1}$ to over 2000 kg ha$^{-1}$. A second advantage of such intensified taungya cultivation is that it requires more labour (around 230 man-days ha$^{-1}$) than the traditional taungya practices, thus contributing to lowering of rural under-employment. Although the intensive cropping practices require relatively high inputs and investments, the agricultural production increases so much, that such intensification is highly profitable to farmers. For instance, in one area it was found that farmers' income (including returns for labour) could increase from Rp 240000 to Rp 915000 per hectare (Sumitro, 1990). In another study it was calculated that crop intensification became profitable to farmers if rice yields were higher than 1700 kg $ha^{-1}$ (Wiersum, 1981).

In order to stimulate farmers to apply the intensive cropping practices, originally the Forest Service not only provided extension services, but also credit to buy the necessary commercial inputs. The distribution of these inputs was centralized by the Forest Service. The farmers had to repay only 70% of the credit for hybrid seeds, fertilizers and insecticides, because it was assumed that 30% of the inputs benefited the tree crops. Indeed, the application of fertilizers significantly increases tree growth (Wiersum, 1981), although data on its long-term effects are still lacking. Further adaptations in the credit scheme have taken place recently. Now the Forest Service only provides credit (free from interest) to farmers for buying hybrid seeds and pesticides. For a few years fertilizers were supplied by the Forest Service provided that the farmers had taken adequate measures for soil and water conservation on the allotted plots. But under the present policy of decreasing government subsidies, taungya farmers now have to buy the fertilizers themselves.

The system for intensified taungya cultivation is now widely practised in all forestry districts. For instance, in the 1989/1990 planting season this system was applied in 74% of the total reforestation area. The objectives of the Forest Service for application of this intensified taungya programme are (Sumitro, 1990):

**1.** to contribute to the national food production;
**2.** to provide more employment opportunities to underemployed farmers;
**3.** to increase the income and welfare of village communities living near forest areas;
**4.** to safeguard forest plantations;
**5.** to maximize returns of the land which is the most scarce production factor in forest areas.

When comparing these objectives with the original ones for the introduction of taungya, it becomes apparent that the priority is no longer only the efficient establishment of timber plantations. Increased attention is being given to the socioeconomic development of local people. It is now recognized that in forest management a balance should be achieved between the needs for local community development and efficient timber production for commercial needs.

## Recent Developments

Although the system of intensified taungya is now widely practised, this has not yet solved all problems related to the high population pressure on the teak forest areas. Consequently, further adjustments of the taungya system are being tried out: increasing the spacing of timber trees and introducing perennial crops into the system.

According to the standard rules for teak forest management the planting distance should be 3 × 1 m. Originally, this spacing allowed a period of 2 years intercropping. But a side-effect of the crop fertilization has been faster tree growth and a somewhat diminished cropping period. It has been suggested that it would be possible to extend the period of intercropping to 4 years by enlarging the planting distance, e.g. to 6 × 1 m. Theoretically, the final timber production would hardly be reduced by such larger planting distances; only early yields of thinnings will be affected. It is also expected that the tree form could be maintained by pruning, while soil protection could be assured by using double instead of single rows of covercrops. Economic calculations (van der Hout, 1984) have indicated that such alternative techniques could increase the net income of farmers by 60%. These adjustments would hardly influence teak production and financial returns to the forest enterprise. The total net value of such an adjusted taungya system would, depending on the interest rate, increase by 25–50% in comparison with the current system. However, the merits of changing the decades-old standard management practices for teak forests are still under discussion.

A second adjustment presently being tried out is in the crops in the taungya system. Originally only annual foodcrops were allowed to be intercropped, but trials are underway to diversify the crop composition by introduction of perennial crops. Not only are understorey crops such as grasses for animal fodder or pineapple used in these experiments, but also small amounts of selected fruit trees such as *Annona* spp., *Psidium guajave*, *Mangifera indica*, *Arthocarpus integra* and *Durio zibethinus*. Obviously, the maintenance of such trees is only possible with increased planting distances of teak (4 × 2 – 6 × 1 m). All fruits and non-timber products from these perennial crops may be collected by the farmers, but the forest enterprise remains the sole owner of all timber, including that of the fruit trees.

This approach is now being tried out in an experimental social forestry programme, which started in 1984. In 1987 around 60 experimental plots of 25 ha each had been established, part of which were in teak forest areas (Bratamihardja, 1987). In this scheme no standard technological packages are used anymore, but the tree mixtures and the planting configurations are decided upon in joint discussions between forest service personnel and the local farmers. In this scheme local Forest Farmer Associations are formed to create a basis for equal partnership between the forest enterprise and local farmers and to facilitate joint decision-making on the locally most suitable plantation mixtures (Kartasubrata, 1990). Thus, this scheme forms one more step in the direction of adjusting the standard forest management practices to the needs and wishes of the local villagers.

## Discussion

During the last decades a gradual intensification of the taungya system on Java has taken place in response to the increasing population pressure on the teak forests. The changes introduced have been aimed specifically at increasing food production and creating more jobs for the local people. In addition to these developments, other efforts have been undertaken since 1974, within the framework of the so-called 'prosperity approach', to improve the living standards of the people living near or in forest areas. A major objective of these efforts was once again to limit illegal cutting and other forms of degradation of the forest resources (Atmosoedardjo and Banyard, 1978; Atmosoedardjo and Wahyudi, 1980). In all these schemes, serious efforts have been made to improve the livelihood and living conditions of the rural people. All these schemes were, however, still within the framework of a conventional forest management approach, in which commercial timber production was considered to be the main priority of the forest enterprise. Although the attitude of the forest service is gradually changing, the various taungya schemes are often still primarily considered as practical tools for the establishment of commercial timber plantation. The subsequent management and utilization of these plantations are still based on the standard management regimes for timber production in which no attention is given to socioeconomic needs of the local population. In this respect it is noteworthy that the intensification efforts of the taungya system were directed at improving food production and employment generation. But in adjusting these systems little attention has been given to possibilities of improving yields of forest products for local subsistence use, such as fuelwood, fodder or construction wood. Therefore it is not surprising that illegal cutting of fuelwood and construction wood for local needs is not stopped by taungya cultivation, nor unregulated livestock grazing in older forests (Simon, 1989).

It is now gradually being recognized that in order to adjust forest management on Java to the prevailing socioeconomic conditions, attention should not only be given to selective elements of the forest management cycle. The total management regime should also be adjusted, bearing in mind not only local requirements for food and employment, but also all forest-related needs of the local population. Such needs are mostly location-specific, i.e. depending on the local population density and ecological conditions. Consequently, it is necessary that specific socio-ecological zones are identified, which are characterized by local requirements for food, employment and forest products for the present and the foreseeable future. Fulfilment of these requirements should be matched with local opportunities for commercial timber production, so that location-specific forest management is achieved, with different degrees of combined commercial and subsistence production. In areas near population centres, the major emphasis should normally be on the production of subsistence needs with only limited commercial timber production. In remote areas commercial timber production could still predominate (Simon, 1990).

In designing management regimes for the various types of location-specific forest management, it will be necessary to give attention to a greater variety of people-oriented management tools than taungya alone. For the various sequential phases of plantation management, different combinations and intensities of combined cropping of timber trees with either food, fodder or fuelwood crops should be developed. The lessons learned from the various efforts to intensify the taungya system should be further applied and extended to include the total forest management regime.

## References

Atmosoedardjo, S. and Banyard, S.G. (1978) The prosperity approach to forest community development on Java. *Commonwealth Forestry Review* 57(2), 89–98.

Atmosoedardjo, S. and Wahyudi (1980) Forest management on Java in the development era. *Nederlands Bosbouw Tijdschrift* 52 (6), 153–165.

Becking, J.H. (1928) *Cultivation of Teak in Java: A Study About Results of Different Regeneration Methods for Teak.* Mededelingen Proefstation v.h. Boschwezen No. 22, Bogor. (Dutch, English summary.)

Beekman, H.A.J.M. (1949) *Silviculture in Indonesia.* Veenman and Zonen, Wageningen, the Netherlands. 386 pp. (Dutch, English summary.)

Bratamihardja, M. (1987) *Social Forestry on State Forest Lands on Java. Paper Workshop Planning and Implementation of Social Forestry Programmes in Indonesia.* Forestry Faculty Gadjah Mada University, Yogyakarta.

Coster, C. and Hardjowasono, M.S. (1935) The influence of agricultural crops in taungya plantation on the growth of teak. *Tectona* 28(6), 464–483. (Dutch, English summary.)

Hardjosoediro, S. (1979) *Forestry Under Conditions of Population Pressure in Indonesia.* Australian National University, Development Studies Monographs 17, 104–115.

Hellinga, G. (1952) Forestry and food production. *Rimba Indonesia* 1 (1), 3–26.

Kartasubrata, J. (1979) Tumpangsari method for establishment of teak plantations in Java. Proceedings of the Symposium on Tropical Agricultural Technologies, Tsukuba, Japan. *Tropical Agricultural Research Reports (Japan)* 12, 141–152.

Kartasubrata, J. (1990) *Lessons from Successful Agrosilvopastoral Projects in Indonesia. Paper Expert Consultation on Agroforestry in the Asia-Pacific Region.* FAO Regional Office for Asia & Pacific, Bangkok, Thailand.

Sabarnudin, M.S. (1988) Effects of agroforestry practice on growth of teak, crop production and soil fertility. PhD thesis, Michigan State University, East Lansing, USA.

Simon, H. (1989) Social forestry as a tool for management of teak forests near heavily populated areas: a case of Java. In Vergara, N.T. and Fernandes R.A. (eds) *Social Forestry in Asia, Factors that Influence Program Implementation.* Regional Center for Graduate Study and Research in Agriculture, Los Banos, Philippines, pp. 193–207.

Simon, H. (1990) An analysis and design of a management strategy for teak cultivation areas to anticipate increasing people's needs. PhD thesis, Forestry Faculty, Gadjah Mada University, Yogyakarta, Indonesia.

Sumitro, A. (1990) *Agroforestry Development in Indonesia. Paper Expert Consultation on Agroforestry in the Asia-Pacific Region.* FAO Regional Office for Asia & Pacific, Bangkok, Thailand.

Van der Hout, P. (1984) Effects of wider initial spacing of teak (*Tectona grandis*) on income and income distribution in the taungya system in Java. *Netherlands Journal of Agricultural Science* 32, 139–142.

Wiersum, K.F. (ed.) (1981) *Observations on Agroforestry on Java.* Forestry Faculty Gadjah Mada University, Yogyakarta, Indonesia & Department Forest Management, Wageningen Agricultural University, Wageningen, the Netherlands, 132 pp.

Wiersum, K.F. (1982) Tree gardening and taungya on Java: examples of agroforestry techniques in the humid tropics. *Agroforestry Systems* 1(1), 53–70.

Wirjodarmodjo, H. and Bratamihardija, M. (1984) Policies, strategies and designs of forest development on the island of Java. In Wiersum, K.F. (ed.) *Strategies and Designs for Afforestation, Reforestation and Tree Planting.* PUDOC, Wageningen, the Netherlands, pp. 363–374.

# Chapter 11
# Taungya in the Philippines

Carl F. Jordan

*Institute of Ecology, University of Georgia, Athens, Georgia 30602, USA*

Deforestation in the Philippines, as in many other developing countries, was initially linked to colonial exploitation, then later to rapidly increasing population, poverty, and inequity of access to forest resources (Sajise, 1991). In 1575, forest covered 92% of the land area of the Philippines. Between then and 1800, only 5% of the forest was removed (Table 11.1). In the 1800s, forest destruction accelerated. Leading causes were shipbuilding, mining and tobacco curing (which requires fuelwood). Between 1900 and 1940, deforestation continued as agriculture increased and log export to China became important. After 1940, mechanized logging and log exports to Japan further decreased native forest (Table 11.1).

Since the Second World War, the rapid increase in population of the Philippines (Table 11.2) has often been cited as the primary cause of deforestation, because more people are clearing forests for agriculture (USAID, 1990). However, the Philippine Government's liberal policy on forest management and regulation has also been a factor, in that it has led to a boom in illegal logging, land speculation and other destructive practices. Deforestation and resource degradation during this period occurred primarily on land considered to be in the public domain. Strong interconnections between the landed elite, the provincial and municipal governments and the business community have often encouraged land use practices that are corrupt, destructive, and inequitable (USAID, 1990).

During the 1950s, logging was carried out primarily by small scale entrepreneurs. As the timber industry became more profitable, the government established a series of regulations that discouraged small-scale legal operations by requiring concessions. The allocation of these concessions was based on political considerations, and the industry came to see timber as a source of immediate wealth rather than a sustainable resource (USAID, 1990).

**Table 11.1.** Decline in forest area in the Philippines, 1575–1990.

| Year | Total forest (ha × $10^6$) | Percentage of total area | Primary forest (ha × $10^6$) |
|---|---|---|---|
| 1575 | 27.5 | 92.0 | |
| 1800 | 26.1 | 87.0 | |
| 1900 | 18.0 | 60.0 | |
| 1934 | 17.0 | 56.7 | 11.1 |
| 1969 | 10.1 | 33.3 | 4.4 |
| 1976 | 8.1 | 27.0 | |
| 1980 | 7.4 | 24.7 | |
| 1988 | 6.5 | 21.5 | |
| 1990 | 6.1 | 20.3 | 0.9 |

Source: Agaloos, 1990 and U.P. National Assessment 1991, cited in Sajise 1991.

**Table 11.2.** Population growth rate in the Philippines, 1948–1990.

| Year | Enumerated population | Average Annual Intercensal Growth Rate (%) |
|---|---|---|
| 1948 | 19234182 | |
| 1960 | 27087685 | 3.06 |
| 1970 | 36684486 | 3.01 |
| 1980 | 48098460 | 2.75 |
| 1990 (estimate) | 62409000 | 2.63 |

Source: Zosa-Ferranil, 1991, cited in Sajise, 1991.

Half the land in the Philippines is reserved in permanent public ownership. When roads are built into forests on this land, shifting cultivators quickly follow (Fig. 11.1). Government control over these lands is very ineffective. As a result of open access, landusers have no incentives to manage for long-term sustainability. When they have no guarantee to land tenure, management is for short-term profit (or survival) rather than long-term sustainability (USAID, 1990).

The impacts of forest removal in the Philippines have been described in the University of Philippines National Assessment (1991) cited by Sajise (1991). They include the following.

**1.** Displacement of cultural communities, especially those who are fully dependent on the forest ecosystem for their existence.

**Fig. 11.1.** Shifting cultivator clearing land in forest recently logged, Claveria, Mindinao, the Philippines, 1991.

**2.** Hydrology impairment such as very high river flows during the wet season and very low flows during the dry season.
**3.** Increased siltation and soil erosion causing declining crops and livestock productivity in the uplands, high maintenance costs for irrigation canals, decreased irrigation and power capacity for infrastructures and decline in fisheries and coastal resource productivity.
**4.** Reduction in biological diversity and species extinction due to habitat destruction.
**5.** Low income, poverty and malnutrition.

Forest removal also leads to the extinction of the timber industry. The forestry sector accounted for 9.6% of the Philippines Gross National Product in 1970, 3.7% in 1980, and 2.5% in 1988 (USAID, 1990).

## Social Forestry

Prior to the 1960s, shifting cultivators that squatted on public lands in the Philippines were subject to incarceration and ejection (Aguilar, 1986). In the early 1970s, the government realized that its punitive approach to the 'slash and burn' (kaingin) problem did not work (USAID, 1990). It was realized that upland and forest occupancy was a social and economic problem that required solutions by the government. As a result, there emerged various socially-orientated forestry projects such as the Forest Occupancy Management Project (FOM), the Family Approach to Reforestation (FAR), the Communal Tree Farming Programme (CTF), Industrial Tree Plantation (ITP), Tree Farming (TF) and Agro-Forestry Farming (AFF). In 1982, these programmes were merged into one Integrated Social Forestry Program (ISFP) (USAID, 1990; Sajise, 1991).

The main objective of these programmes was to keep the upland farmer within the piece of forest land that he occupied, with minimum soil erosion and environmental degradation, and with the long term prospect of bringing back a forest cover in the area. Under the ISFP, forest occupants are granted a Certificate of Stewardship Contract for 3–7 ha, with a 25-year lease that is renewable for another 25 years. For leases to be renewed, land must be cultivated with conservation practices, and 25% must be in reforestation.

As a national programme designed to slow down the rate of deforestation by containing the activities of upland occupants, the ISFP has not seemed to be significantly effective. Reasons suggested by Sajise (1991) include insufficient and inadequately trained personnel, and shortage of funds. However, doubts on the part of the peasants about land security and land adequacy also played an important role. The general trend is for upland cultivators to want secure land title before they begin to invest efforts in soil conservation and reforestation (Aguilar, 1986).

## Plantation Forestry

Industrial forest plantations are considered to be the best hope for timber sources in the future for the Philippines (USAID, 1990). In 1990, only about 466000 ha were covered by forest plantations. However, the Master Plan for Forestry, part of the strategy for sustainable development of natural resources of the Government of the Philippines, includes a greatly expanded forest management programme, emphasizing industrial plantation forests.

Whether any of the existing or planned forest plantations are taungya forestry, in the sense used in this book, is not clear. In general overviews of agroforestry (e.g. MacDicken, 1990), it is sometimes claimed that 'taungya' is carried out in the Philippines. In the sense that taungya is a system of forest plantations in which peasants cultivate crops for the first few years between the seedlings of a forest plantation, this is correct. The Paper Industries Corporation of the Philippines joined with the Philippines Development Bank in a programme that provides loans, technical assistance, and a guaranteed market to individual landowners who produce pulpwood for nearby pulp mills (Kirchhofer and Mercer, 1986). Figure 11.2 illustrates the plot of a private landowner in Mindanao, Philippines, in which upland rice is being grown within a plantation of *Gmelina arborea*

**Fig. 11.2.** Upland rice interplanted in a plantation of *Gmelina arborea* in Mindanao, the Philippines, 1991.

destined for a local pulp mill. Gmelina is a fast-growing tree that can reach a harvestable size within 6–8 years, and yield a profit to the private landowner within a time-span that is well within his economic horizon.

However, this programme does not result in reforestation of slower-growing, high-value timber. This is unfortunate. The natural dipterocarp forests of the Philippines have been extremely important for the national economy over the past decades, but as mentioned above, their importance for the national economy has dropped from about 10% to about 2% of the gross national product over the past 20 years. The high-value native hardwood forests are not being re-established, for technical and economic reasons.

Technically, it is more difficult to establish plantations of native hardwoods than of exotics such as eucalyptus and acacia. Seed formation of native species often is erratic and seed trees dispersed, so that a continuing major effort must be made to collect sufficient seeds. Many tropical hardwoods have seeds which remain viable for only several weeks after ripening, and this makes seed collection and storage difficult. Seeds are not genetically uniform, and so progeny may have various shapes and forms, undesirable from the forester's point of view. Seedling establishment requires special care. Many species may not grow well in the direct sunlight that is characteristic of young plantation forests. Nurse trees may be necessary to improve light conditions, microclimate, and soil characteristics. Often, exotics such as eucalyptus are supplied with fertilizers to ensure rapid early growth, but wild species may not respond well to fertilizers. Soil microfauna and mycorrhiza that are essential for some native species may be destroyed by land-clearing techniques often used in commercial plantations. For these reasons, foresters often prefer to work with common plantation species such as eucalyptus which are better suited to the artificial, human-managed conditions of traditional plantations.

Plantations of exotics can be desirable as long as capital is available to manage them. But long-term survival of plantation forests where capital inputs may be sporadic, as in developing countries, may be better achieved with native species. Native species are better adapted to local soil conditions, and can be more resistant to local pests. Most important, they can survive the competition with local 'weed' species, whereas weeding is often critical in exotic plantations.

Economic factors are another reason that small landowners favour fast-growing species such as gmelina. They can plant, harvest, and reap an economic income within a time-span of less than 10 years, a reasonable economic horizon. Native dipterocarps could take 50 years to reach a marketable size, and individuals of the native forests that bring the most profit may be 100 years old or older.

## Taungya

It can be argued that the only way to re-establish high value native species in forest plantations would be through modified taungya in the sense that it is used in the Forest Village System of taungya in Thailand. It may be that only through this approach can the economic, social, and technical barriers to re-establishment of native species be overcome.

### Economic

The length of time required for most native species such as dipterocarps to reach maturity and become ready for harvest is so long that an individual landowner would not be likely to invest. Investment would have to come from the government, international banks, or development agencies.

### Technical

Despite the technical difficulties associated with establishing a nursery for native seedlings, and establishing the species in plantations, it can be done. It simply is more difficult and expensive than working with traditional exotics, at least at present. This need not always be so. An initial investment is needed to develop the techniques and establish the infrastructure that would make native forests competitive economically with exotics.

Another obstacle to be overcome in re-establishing native forests is the preference of foresters for planting monocultures. Planting and harvesting polycultures is more complicated than planting and harvesting monocultures. However, planting and harvesting costs must be weighed against benefits of polycultures, such as decreased maintenance costs, improved hydrology, and more diversified habitat leading to conservation of species diversity.

### Social

Social problems would be the most difficult to overcome, to establish taungya for native species in the Philippines. They are basically the same as social problems facing taungya in other countries, including the peasants' desire for land ownership and the lure of higher wages in the cities. Higher wages is a surmountable problem, if long-term investment is available. The land tenure problem is the most difficult.

In Philippine social forestry programmes, it is virtually impossible for authorities to stop poor stewardship and inappropriate land management, before damage is done. The only enforcement tool available is the threat of non-renewal of the lease. However if the end of the lease is many years in the future, the careless peasant hopes that government regulations will change by the time the lease is up, or that authorities can be persuaded to overlook infractions.

In contrast, in taungya, managers have much more control over tree plantations. It is much more difficult for peasants to succeed in illegally cutting trees within a system of taungya. This is not to say that social considerations are unimportant within taungya. But conditions need be only slightly better than conditions elsewhere, to persuade peasants to voluntarily remain in the system. It would be nice to give land tenure to peasants, but it is not essential. After all, migrants who leave for jobs in the city give up all hope of getting land tenure.

### Culture

Success in establishing taungya may be related to cultural tradition (Jordan, 1990). A strong, hierarchical, authoritarian tradition of government as exists in Thailand, and a religion such as Buddhism which places less importance on material acquisitions, may provide a setting in which taungya may be more easily facilitated. In contrast, in the Philippines with a tradition of *laissez faire* capitalism, and Christianity which is often interpreted as putting man above nature (White, 1967), taungya may be more difficult.

## Conclusion

Re-establishing native dipterocarp forests may be one of the most important activities that can take place in the Philippines. It is important for conservation, because it would help protect watersheds and help preserve biological diversity. It is important economically, because it has the potential to generate future income.

Re-establishing native species will probably not occur by means of community and social forestry programmes, because time to pay-off is too long. Native dipterocarp forests could be re-established through methods of traditional industrial forestry, modified to accommodate requirements of native species mentioned above. However, industrial forestry faces competition for land by badly needed agriculture. Taungya is one approach that can help lessen this competition.

Even with taungya, need for economic income from a piece of land during the period of time before the trees are harvested presents a problem. Forests in which high value hardwoods are mixed with fruit trees that give income for the first decade or two of a plantation may be a partial solution.

Benefits and problems of this approach have been discussed in detail in Chapter 7.

## References

Aguilar, F.V. (1986) Finding from eight case studies of social forestry projects in the Philippines. In Fujisaka, S., Sajise, P.E. and Del Castillo, R.A. (eds) *Man, Agriculture and the Tropical Forest: Change and Development in the Philippine Uplands.* Winrock International Institute for Agricultural Development. Bangkok, Thailand, pp. 323–335.

Jordan, C.F. (1990) Why Agroforestry Systems work better in Asia than in South America. *Symposium on Tropical Forest Ecology, The Changing Human Niche and Deforestation,* Annual Meeting of the American Anthropological Association, New Orleans, LA. Nov., 1990.

Kirchhofer, J. and Mercer, E. (1986) Putting social and community forestry in perspective in the Asia-Pacific region. In Fujisaka, S., Sajise, P.E. and del Castillo, R.A. (eds) *Man, Agriculture and the Tropical Forest: Change and Development in the Philippine Uplands.* Winrock International Institute for Agricultural Development, Bangkok, Thailand, pp. 189–222.

MacDicken, K.G. (1990) Agroforestry management in the humid tropics. In MacDicken, K.G. and Vergara, N.T. (eds) *Agroforestry: Classification and Management.* Wiley, New York, pp. 98–149.

Sajise, P.E. (1991) *The evolution of social forestry in the Philippines: lessons learned and some questions asked.* Lecture presented at Wageningen Agricultural University, Dept. of Forestry, The Netherlands, June 19, 1991.

USAID (1990) *Natural Resources Management Program (492–0444)* Program Assistance Approval Document. U.S. Agency for International Development/ Philippines.

White, L. (1967) The historical roots of our ecologic crisis. *Science* 155, 1203–1207.

Chapter 12

# Taungya in Nepal

Jay B.S. Karki[1] and Binod Bhatta[2]

*[1]Social Science Division, International Rice Research Institute, PO Box 933, Los Baños, Manila, Philippines;*
*[2]Ministry of Forestry, Kathmandu, Nepal*

The degradation of land in the hills and mountains of Northern Nepal in the past decades has resulted in migration of landless people to lowland forest areas of Nepal. As a result of their clearing for agriculture, there has been massive deforestation which has resulted in a shortage of wood for fuel. Ninety five percent of the people of Nepal depend on wood for an energy supply. Continued deforestation is reducing even further the available supply of fuelwood. Natural regeneration and forest growth is not capable of keeping up with the demand. Nepal faces urgent problems in solving agricultural problems, in feeding the displaced citizens, and in obtaining fuelwood for energy.

In 1976, His Majesty's Government of Nepal prepared a National Forestry Plan which emphasized the importance of improving forest productivity and the need for community forestry programmes to meet local demands for fuelwood, timber, fodder and other forest products.

The Tamagadi Project (Fig. 12.1) was started in the early 1980s, with settlers being mostly hill migrants. Initially the project had 120 families participating, but it now has 400 families in four settlement sites. The Tamagadi Project began with no clear rules and regulations, no detailed plan for reforestation, no criteria for selection of farmers, and no scheduling of technical activities. The project operated independently until 1984/85.

When reforestation by plantations began, taungya was seen as an effective means of accomplishing the task. However, as more and more immigrants entered the area in need of agricultural land, there was a need to adjust taungya so that needs of rural development as well as reforestation were served. There was a need to have longer periods of cropping, wider distances between planted trees, and/or shorter tree rotations. Based

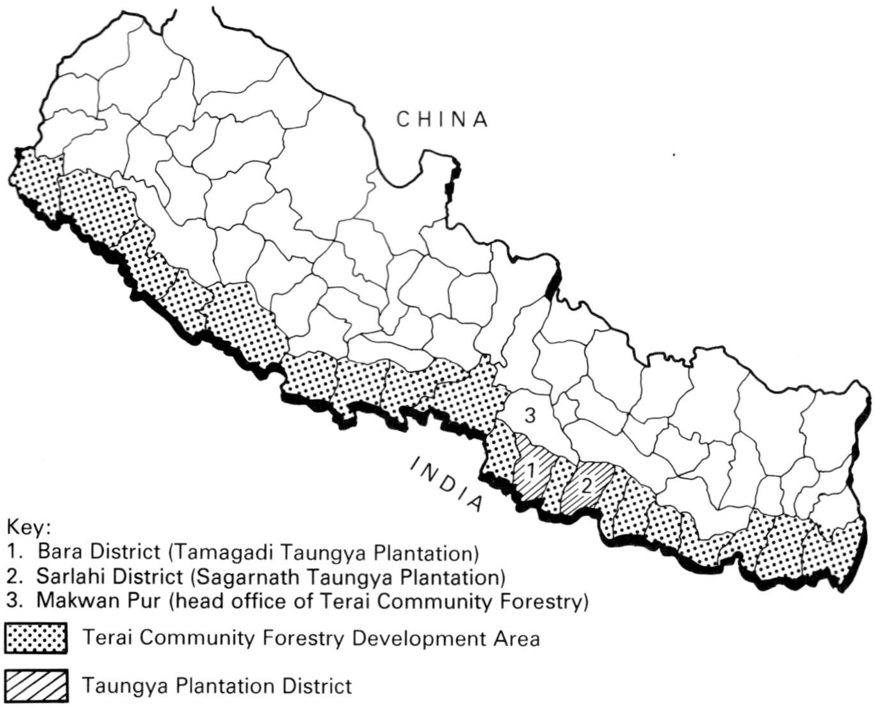


**Fig. 12.1.** Map of Nepal, showing the locations of community forest projects and taungya plantations.

on the lessons of this project, the Terai Community Forestry Development Project and the Sagarnath Forest Development Project within the Sarlahi District (Fig. 12.1) were initiated with the support of the Asian Development Bank and the World Bank.

The Sagarnath Project is unique in Nepal, both in its size and scope. It has an annual programme of clearcut harvesting of 1000 ha of depleted forest, and the entire project covers an area of around 16000 ha. The portion of the forest that is cut is replanted during the same year as the harvest. Establishment requires site preparation and planting of some 1500000 seedlings, as well as weeding and fire protection (White, 1986). As a result of this activity, there are organizational strains in implementing this diverse programme.

Because of the need for high forest yield, part of the plan was to convert natural forest to plantation. Where other forest values had public support and demand, the native forest was retained in parks and reserves (White, 1986).

## Taungya Plantation System in Practice

The first step in converting low value forest lands to taungya is clearcut felling. Trees are removed by the Timber Corporation of Nepal and the Fuelwood Corporation of Nepal. After harvest, good all-weather roads are constructed around blocks 1000 m on a side. Within the blocks are 6-m-wide access roads/firebreak strips.

Site preparation is carried out with D6 caterpillar tractors. Then the farmers plough the land and dig the pits for planting of tree seedlings. Planting is done by a team of about 30 workers. Seedlings are provided free to the farmers. Farmers are paid for their labour only during the first year, for site preparation and planting of seedlings.

After planting, the land is allocated to the farmers for intercropping of agricultural crops. Intercropping not only gives the farmers an income, it also controls the weeds which otherwise would compete with the tree seedlings. In addition to the weeding benefit, fire control is also an important reason for interplanting. Cultivation of crops is permitted for the first 3–4 years of the plantation.

The farmers prefer traditional grain crops such as upland rice and corn for the rainy season, and oil seed (rape) in the winter. These crops are not planted until the tree seedlings are large enough so they are not damaged by competition. There have been a wide variety of other crops used for intercropping (Table 12.1) (White, 1985). Horticultural crops have also

**Table 12.1.** Agricultural crops: season and plantation time cycle.

| Year | Season | Crop |
|---|---|---|
| Prior to tree planting | Rainy | Maize, dryland rice, chilli, groundnut, soya, sesame (til), niger |
| | Dry | Oilseed rape (toria), linseed, safflower, tomato, potato, wheat, millet, pulses, tobacco |
| First plantation year | Rainy | As above except sesame |
| | Dry | As above except sesame |
| Second plantation year | Rainy | Maize, soya, pigeon pea (rahar, Cajanus) sesame, niger |
| | Dry | Oilseed rape, safflower, linseed, sesame, pulses, wheat |
| Third plantation year | Rainy | Maize, as a fodder crop only |
| | Dry | Sesame – a possibility |

Note: soya, sesame, tobacco and niger are planted late in the rains, ripening in winter.

proven feasible (Table 12.2), and there have been successful trials of combinations of agricultural, horticultural, and medicinal crops (Table 12.3). There is little use of commercial fertilizers or farmyard manure in the fields, because of the cost and the problem of transportation. Despite the lack of fertilizer, crop production is generally satisfactory.

The commonest tree species planted in the Tamagadi Project are sisso (*Dalbergia sisso*) and teak (*Tectona grandis*). About 19 species were evaluated for forest plantations in the Sagarnath project.

Arrangements with local cultivators differ between the two projects. In Sagarnath, the cultivators are chosen from the surrounding villages. The farmers remain in the villages, and do not move to the project site. In Tamagadi, farmers live on the site, and must move after 4 years. Therefore, they build temporary housing. At both sites, there are supervisors that have responsibilities to oversee the site and ensure proper growth of the plantations. They also give advice to the farmers about crop cultivation.

## Evaluation of the System

The taungya system has reduced substantially the costs of reforestation, as well as giving an alternative way to cultivate cash and food crops to the poor and landless farmers, and providing them with an income (White, 1986). The system also has reduced the incidence of fire, further increasing the economic benefit of the system.

**Table 12.2.** Suitability of soil types for horticultural and medicinal crops.

| Crop | Soil type | |
|---|---|---|
| | Sandy loam | Silty clay loam |
| Banana | – | * |
| Ginger | ? | * |
| Papaya | – | * |
| Pineapple | ? | * |
| Rauwolfia | * | * |
| Turmeric (besar) | ? | * |

Source: White, 1986.
*, suitable; ?, anticipated as suitable; –, not suitable.

**Table 12.3.** Horticultural shrub crops – spacing and yield.

| Crop | Tree line spacing | Crop lines | Crop space (m) | No. of crop plants | Yield at 70% production | | Field value Rs |
|---|---|---|---|---|---|---|---|
| Bananas | 4 × 2 | 4× | 2.0 | 1250 | 875 bunches | 200 | 22 000 |
| Bananas | 4 × 2 | 2× | 2.0 | 2500 | 1750 bunches | fruit/bunch | 44 000 |
| Bananas | 5 × 2 | 2× | 1.5 | 3000 | 2100 bunches | @8 Rs 1/– | 52 000 |
| Papaya | 4 × 2 | 4× | 2.0 | 1250 | 875 trees | 10 | 6 500 |
| Papaya | 4 × 2 | 2× | 2.0 | 2500 | 1750 trees | fruit trees, | 13 000 |
| Papaya | 5 × 2 | 2× | 1.5 | 3000 | 2100 trees | @0.75 Rs | 15 750 |
| Pineapples | 4 × 2 | 0.5× | 0.5 | 25000 | 17500 fruits | @1 Rs 1/– | 17 500 |
| Pineapples | 5 × 2 | 7× | 0.5 | 28000 | 19600 fruits | | 19 600 |

Source: White, 1986.
Note: Values, banana and papaya are 50% of local market price, pineapples 25%. Kathmandu prices much higher.

## Environmental Effects

Forestry plantations, whether established by taungya or by strict plantations, change the landscape. The original landscape, although not valuable for wood products, is habitat for wildlife. Establishment of plantations has in some cases, affected the population of tigers and elephants. However, destruction of wildlife habitat by plantations must be weighed against the destruction that occurs when peasants scavenge the forest to obtain fuelwood and building materials. In plantations, the system is at least sustainable, while scavenging eventually destroys the system.

A further consideration is that forest plantations have resulted in the increase in some mammals such as deer, and birds such as the red jungle fowl, the common quail, the peafowl and parakeets.

## Policy Implications

The welfare of the farmer is a key factor in the success of the programmes. Unless the problems of the peasant farmers are resolved, there can be no resolution to the problem of reforestation and wood supply.

There must be close cooperation between the project staff representing the owner of the forest land, and the labourers who work there for their subsistence. The project must be coordinated also with other agencies concerned with providing basic developmental support. Relevant knowledge and skills should be imparted to the farmers through training and extension. The training programme should also incorporate environmental education, family planning, welfare, supportive livelihood technology, and skills.

Specific recommendations regarding taungya have resulted from the experience at Sagarnath. The spacing between plantation seedlings should be increased to 5 × 2 m, so that there is less competition between trees and crops. Wider spacing increases the length of time during which interplanting is possible. Intercropping should be allowed for longer, as it helps to control weeds when the farmer tills the land. After 3–4 years, when productivity of agricultural crops decreases, the area should be intercropped with some shade-tolerant horticultural crops and medicinal crops. Grasses and other fodder for cattle grow in the land between the trees after cropping has ceased.

In general, any measures which can increase and prolong crop production and thereby benefit the peasants are desirable. Such measures may not increase reforestation as fast as a system of pure forestry. However, it will ensure that the whole system, including the social system of the rural countryside is more stable, and in the long run, both the forests and the peasants will benefit.

## References

White, K.J. (1985) *Intercrop Strategies and Practices in the Bhabar Terai of Central Nepal (manual # 1)*. Ministry of Forests, Kathmandu, Nepal.

White, K.J. (1986) *Tree Farming Practices in the Bhabar Terai of Central Nepal (manual # 2)*. Ministry of Forests, Kathmandu, Nepal.

# Chapter 13
# Taungya in India

K.P. Singh[1] and O.N. Pandey[2]

*[1]Department of Botany, Banaras Hindu University, Varanasi, India; [2]Faculty of Forestry, Bisra Agricultural University, Ranchi, India*

Taungya was begun in India in the late 19th century, as a means to exploit the cheap labour of landless peasants in establishing plantations of economically valuable trees, especially Sal (*Shorea robusta*) and teak (*Tectona grandis*).

Today, taungya is practised in several regions of India (Table 13.1). The land is owned by the State Forest Department, and the forest dwellers (temporary or permanent) carry out various planting operations in return for crop yields and other social benefits. Three different systems have been tried.

**1.** The Department pays daily wages to peasants for their labour in planting trees. This system has been least successful.
**2.** The land is leased to the peasants for 2–3 years for cropping.
**3.** The village system of taungya, which appears to be the most effective.

Sal (*Shorea robusta*) is the main plantation species in taungya. In the Gorakhpur Forest Division, the total forest area is 55 800 ha, of which 20 000 ha is in taungya. In the plantations, Sal comprises 82% of the stock, teak 4% and other species 14%. However, other tree species are often included (Table 13.1).

The crops which are intercropped between the trees are also presented in Table 13.1. Some of the agricultural crops are monsoon crops (kharif) while others are winter crops (rabi) (Table 13.2). The schedule for planting and cultivation for the first years of taungya in the Gorakhpur Forest Division is given in Table 13.3. Site preparation includes felling of original trees and their removal, and construction of trenches to protect plantations from cattle. The Kharif crop during the first year is often maize or paddy rice, and the Rabi crop is often wheat or mustard. The second year, the

**Table 13.1.** Crops grown in taungya in India.

| State | Tree crop | Associated agricultural crops |
|---|---|---|
| Uttar Pradesh | *Shorea robusta*<br>*Tectona grandis*<br>*Acacia catechu*<br>*Dalbergia sisoo*<br>*Eucalyptus* spp.<br>*Populus* spp. | Maize, paddy, sorghum, pigeon pea, soybean, wheat, barley, chickpea, oilseed rape, etc. |
| Andhra Pradesh | *Anacardium occidentale*<br>*Tectona grandis*<br>*Bombax ceiba*<br>Bamboo<br>*Eucalyptus* spp. | Hill paddy, groundnut, sweet potato |
| Kerala | *Tectona grandis*<br>*Bombax ceiba*<br>*Eucalyptus* spp. | Paddy, *Manihot* sp. ginger, turmeric |
| Assam | *Shorea robusta* | Paddy |
| Tamil Nadu | *Tectona grandis*<br>Bamboo<br>*Santalum album*<br>*Tamarindus indica*<br>*Acacia nilotica*<br>*Acacia mearnsil*<br>*Ceiba pentandra*<br>Cashew<br>Rubber | Millets, pulses, groundnut, cotton, *Manihot*, potato |
| West Bengal | *Tectona grandis*<br>*Shorea robusta*<br>*Schima wallichii*<br>*Cryptomeria japonica*<br>*Quercus* spp.<br>*Michelia doltsopa* | Paddy, maize, millets, turmeric, ginger, ladies fingers, pineapple, hemp |
| Maharastra | | Sunhemp, jute, mesta, sunflower, castor |
| Andaman and Nicobar Islands | *Pterocarpus dalbergoides* | Sugarcane, maize |

Source: Tejwani, 1987.

**Table 13.2.** Crops interplanted in taungya in three regions of India.

| Region | Monsoon crop (kharif) | Winter crop (rabi) |
|---|---|---|
| Uttar Pradesh | Maize, millet, pigeon pea, cotton, paddy, hemp | Wheat, barley, pea, oilseed rape, linseed |
| W. Bengal | Paddy, millet, turmeric, jute, cotton, sesame, pigeon pea, maize | Various pulses |
| Assam | Ahu, paddy | |

Kharif crop commonly is pigeon pea (*Cajanus cajan*). For years 3–5, the kharif and rabi crops are the same as for year 2. After the fifth year, cropping ends and the villagers move.

In the modern taungya village system, the following facilities are provided free to participating families:

1. timber and thatching material for hut;
2. crop yield (grain plus straw);
3. firewood collection, charcoal making;
4. grazing (1 buffalo and 2 cows per family);
5. primary education (one school per centre);
6. irrigation pump (one per 20 ha);
7. drinking water hand pump (10 pumps/centre).

The economic benefits derived by the villagers from the taungya system, and the value of the wood after 5 years and 40 years are given in Table 13.4. Even after 5 years, the operation can be profitable.

From silvicultural and agronomic aspects, the system of taungya works well. Productivity of grains drops off by the fifth year, but is still significant. However, social and political problems are creating difficulty for the system.

1. There is a reluctance of the peasants to move after the 5-year period of cultivation is over. They would prefer to stay in one location.
2. Some workers are leaving for jobs in industry, creating a labour shortage in the village system.
3. Sometimes non-residents begin cultivation within the village system, and there is land encroachment.
4. There are political problems in the administration of the villages.

Increasing pressures on the land for short-term food production is detracting from the success of taungya, where the main goal is reforestation

**Table 13.3.** The schedule for planting and cultivation of taungya.

| | Year 1 | | | | | | | | | | | | Year 2 | | | | | | | | | | | | Year 3 | | | |
|---|---|---|---|---|---|---|---|---|---|---|---|---|---|---|---|---|---|---|---|---|---|---|---|---|---|---|---|---|
| Month | J | F | M | A | M | J | J | A | S | O | N | D | J | F | M | A | M | J | J | A | S | O | N | D | J | F | M | A |
| *Activities* | | | | | | | | | | | | | | | | | | | | | | | | | | | | |
| Site preparation | — | — | — | — | — | | | | | | | | | | | | | | | | | | | | | | | |
| Kharif crop | | | | | | — | — | — | — | — | | | | | | | | | | | | | | | | | | |
| Rabi crop | | | | | | | | | | | — | — | — | — | — | — | | | | | | | | | | | | |
| Plant sal | | | | | | | | | | | | | | | | | | — | | | | | | | | | | |
| Kharif crop | | | | | | | | | | | | | | | | | | — | — | — | — | | | | | | | |
| Weeding | | | | | | | | | | | | | | | | | | | – | – | – | – | | | | | | |
| Rabi crop | | | | | | | | | | | | | | | | | | | | | | | | — | — | — | — | — |
| Thinning | | | | | | | | | | | | | | | | | | | | | | | | — | — | — | — | — |

**Table 13.4.** Economic analysis of the first 5 years of taungya in the village system, based on 100 families per centre, and 2.5 ha land per family.

| | | Rupees/ha |
|---|---|---|
| *Benefits to villagers* | | |
| Land lease rent subsidy | | 15000 |
| Timber, thatch, grazing, school | | 3050 |
| Food crop, fuelwood | | 22000 |
| Irrigation, drinking water | | 1550 |
| | **Total** | 41500 |
| *Value of wood produced* | | |
| Sal forest wood at 5 years (for poles) | | 700000 |
| Sal forest wood at 40 years (timber) | | 4800000 |

for timber supply. Improvements in the system are required if timber production is to be continued.

## Reference

Tejwani, K.G. (1987) Agroforestry practices and research in India. In Gholz, H.L. (ed.) *Agroforestry: Realities, Possibilities and Potentials.* Martinus Nijhoff, Dordrecht, pp. 109–136.

# Chapter 14
# Taungya in China

Yingqiang Huo

*Forest Research Institute of Guangdong, Guangzhou, China*

Agroforestry is an integrated system for multiple uses of land. Often, agroforestry systems are composed of trees and agricultural crops, but animal husbandry and fisheries also can be included. Taungya is a type of agroforestry system which combines agriculture with commercially valuable timber species. There are many variations of taungya in China, but they fall into three general patterns with each pattern predominating in a different region. In this chapter, the three major patterns of taungya in China are described.

## Pattern I. Trees + Crops in the North China Plain

The land in the lower reaches of the Yellow River was created by silt from the Yellow, Huai, and Hai Rivers, which for most of their course flow over terrain less than 50 m above sea level. The alluvial deposits which cover the area are highly fertile. Annual precipitation that averages about 600 mm, and hot wet summers are conducive to growth, and make the North China Plain an important agricultural area.

### Tree species

The natural vegetation of this region is temperate deciduous broad-leaved forest similar to those in Western Europe, northern Japan, and eastern North America. Since the plain is the country's most important farming area, there are few native forests still in existence. However, for taungya agriculture, commercially valuable timber species such as poplar, willow, elm, Chinese scholartree, black locust, toon (*Ailanthus altissima*), paulownia and catalpas are planted. Deciduous fruit trees such as apple, pear, peach and persimmon also are widely cultivated.

For taungya where crops are planted between or under the trees, the trees must be deciduous, shade-intolerant, and with a small or open crown. The trees themselves are important for harvest in Chinese taungya. They must grow fast, and be of good timber quality. Paulownia is frequently favoured by farmers because of its rooting habit. Their roots are concentrated in the deeper soil. About 80% of their roots are between 40 and 100 cm, and only 8% are above 20 cm. This rooting habit makes paulownia a good tree with which to intercrop agricultural crops, since most crops root at depths less than 40 cm. There appears to be little competition between paulownia and crops for nutrients and water. Paulownia is also desirable because the species has an open crown through which a large proportion of incoming light can penetrate (Table 14.1). Paulownia in China forms its leaves relatively late, in late April to early May, and thus summer-harvested crops can get a good start. Paulownia loses its leaves in mid-November, and its mulch seems to help prevent frost damage to autumn crops.

## Spacing and direction of tree rows

The planting density of paulownia depends upon the purpose of the plantation. For strict forest plantations, trees should be planted at intervals of 5 × 5 m or 6 × 6 m. In taungya, where crops are interplanted between the

**Table 14.1.** Proportion of sunlight penetrating canopies of various taungya tree species.

| Tree species | Times measured | Radiation intensity (cal $cm^{-1}min^{-1}$) | | | |
|---|---|---|---|---|---|
| | | Penetrating patches | Shading patches | Average | %[1] |
| Paulownia | 8 | 0.63 | 0.22 | 0.28 | 100 |
| Willow | 8 | 0.47 | 0.20 | 0.25 | 89 |
| Paulownia | 10 | 0.62 | 0.20 | 0.29 | 100 |
| Chinese toon | 10 | 0.29 | 0.20 | 0.21 | 73 |
| Black locust | 10 | 0.32 | 0.17 | 0.18 | 63 |
| Paulownia | 6 | 0.48 | 0.14 | 0.29 | 100 |
| Populus | 6 | 0.37 | 0.16 | 0.21 | 63 |

[1]% of value for Paulownia.

paulownia, the trees should be planted in rows 20–30 m apart, and with 3–6 m intervals within the rows. This spacing gives the desirable microclimate when the trees are young. After 5–6 years, alternate rows of trees are harvested, leaving the final rows 40–60 m apart.

The direction of rows in which the trees are planted is important in affecting crop yield. Rows running north–south are better than those from east–west, because the crops obtain more light. In north–south rows, the time of minimum shade on the crops coincides with the time of maximum sunlight (noon). In the east–west rows, there is more shade on the crops at noon time, and less in the early morning and late evening than in the north–south rows. The influence of trees on crop yield is evident in Table 14.2, which shows that crops within 1 m of the trees have reduced growth.

Pruning of lower branches of paulownia improves crop yield, apparently because this practice increases the light reaching the crop. Table 14.3 illustrates the influence of pruning lower branches on the yield of corn and sorghum in taungya with paulownia.

**Table 14.2.** Growth and yield of corn and cotton as a function of distance from paulownia planted in alleys.

| Crop | Paulownia | | | Distance (m) | Ht of crop (cm) | Wt/1000 seeds (g) | Yield (kg ha$^{-1}$) | No. of boll |
|---|---|---|---|---|---|---|---|---|
| | Age | Spacing (m) | Ht (m) | | | | | |
| Corn | 6 | 5 × 50 | 5.7 | 1 | 193 | 248.9 | 5303 | – |
| | | | | 9 | 210 | 248.3 | 7553 | – |
| | | | | 19 | 193 | 261.3 | 6953 | – |
| Cotton | 6 | 5 × 50 | 5.7 | 1 | 49 | – | – | 33 |
| | | | | 9 | 78 | – | – | 68 |
| | | | | 19 | 76 | – | – | 60 |

**Table 14.3.** The influence of height of pruning on the yield of corn and sorghum.

| Crops | Ht below limb (m) | Wt per 1000 seeds (g) | Yield (kg ha$^{-1}$) | %[1] |
|---|---|---|---|---|
| Corn | 6.0 | 215.3 | 3270 | 100.0 |
| | 2.3 | 213.0 | 1875 | 57.3 |
| Sorghum | 3.3 | 31.4 | 2722 | 100.0 |
| | 2.3 | 32.7 | 1950 | 71.6 |

[1]% of value of maximum pruning.

## Selection of crops

In order to utilize most fully the environmental resources, and to maximize economic benefits, crops are selected for interplanting or underplanting beneath the trees in accordance with site conditions and the season. Major crops in the North China Plain are wheat, corn, cotton, sesame, peanut and tobacco. Intensity and direction of light are among the first factors considered in planting taungya. Planting density and direction are frequently selected to favour summer crops rather than autumn crops, since yields from summer crops frequently are higher.

Many medicinal herbs and some kinds of vegetables such as celery, ginger, and day lily are partially shade-tolerant, and have higher yield in partial shade. These species may be planted closer to the paulownia, or in plantations where paulownia is spaced more closely.

## Benefits of the pattern

Interplanting agricultural crops between rows of paulownia increased yields on summer harvested crops (Table 14.4) as well as some autumn harvested crops (Table 14.5) compared to controls. Several mechanisms can be suggested. Hot dry winds are common in summer on the North China Plain. The rows of trees may protect the wheat to some extent from the direct effects of the wind. The light shade cast by paulownia also appears to improve the moisture relations in the soil. In the alley systems, the soil moisture is generally higher than in wheat monocultures.

The nutrient content of the leaves and flowers of paulownia also can benefit the growth of interplanted wheat. In an experiment, an addition of

**Table 14.4.** A comparison of yields between interplanted fields and controls of summer-harvested wheat.

| Paulownia | | | Yield ($kg\ ha^{-1}$) | | | Wt per 1000 seeds (g) | | |
|---|---|---|---|---|---|---|---|---|
| Age | Spacing (m) | DBH[1] (cm) | Interplanted | Control | % | Interplanted | Control | % |
| 8 | 10 × 20 | 38.1 | 4320 | 3452 | +19.8 | 37.9 | 36.8 | +3.0 |
| 8 | 10 × 15 | 28.0 | 1233 | 1183 | + 4.4 | 25.6 | 28.2 | −7.0 |
| 10 | 7 × 20 | 34.0 | 3263 | 2543 | +23.0 | 35.5 | 37.7 | −6.0 |
| 10 | 10 × 10 | 40.0 | 3471 | 3142 | +10.0 | 39.4 | 40.2 | −3.0 |

[1]DBH: diameter at breast height.

**Table 14.5.** A comparison of yields between interplanted fields and controls of autumn-harvested crops.

| Crop | Paulownia | | Yield (kg ha$^{-1}$) | | |
|---|---|---|---|---|---|
| | Age | Spacing (m) | Interplanted | Control | % |
| Corn | 8 | 9 × 18 | 2423 | 2184 | + 9.9 |
| Millet | 5 | 9 × 18 | 949 | 788 | +20.0 |
| Cotton | 5 | 8 × 60 | 2678 | 2220 | +17.0 |
| Soybean | 8 | 10 × 15 | 1146 | 1258 | − 8.9 |
| Sweet potato | 10 | 10 × 15 | 12585 | 20880 | −39.8 |

mulch from paulownia leaves increased wheat production by almost 20% over the control.

## Pattern II. Taxodium + Crop + Fishery in Wetlands of the Huai Plain and the Middle–Lower Yangtze Plain

The land from the southern part of the North China Plain down to the Yangtze Delta between Jiangsu and Zhejiang Provinces is mostly less than 10 m above sea level. The region is crisscrossed by rivers and studded by lakes. These bodies of water are a rich source of fish, water chestnuts, lotus seeds and roots, and reeds. The Yangtze Delta is one of China's most important farming areas for rice, cotton, wheat, rapeseed, and ramie. This region of rivers, lakes, and estuaries abounds in rice paddies, mulberry orchards, and fish ponds.

Much experimental work has been done on the lakeside plain of Lake Hongze, in the western part of Jiangsu Province. It has an area of 2500 km$^2$, and is China's third biggest freshwater lake in area, but has an average depth of less than 4 m. Silt deposition from rivers draining into the lake has filled it to the point where it has spilled over during the rainy season, and dried up during droughts.

The wetlands in this region have not been fully exploited because of various environmental problems. The soil has a high clay content, and although the pH is 6.0–7.4, and the organic matter and nitrogen content of the soil is high, the phosphorus content is low, and there is a severe problem of phosphorus deficiency. To make much of the wetlands suitable for agroforestry, some areas must be deepened to make the water deep enough for fish, and other areas must be built up to be high enough for growing crops.

## Types of wetlands

Wetlands can be classified into three major types, according to the fluctuation of the water table.

**1.** Moistland. The water table fluctuates from the surface down to 30–40 cm in the ground. During heavy rains, the surface may be flooded. Moistlands are commonly located along river banks, lake shores, and sea beaches. The vegetation is moist-tolerant and hydrophilous trees, shrubs and herbs. To establish agriculture, the water table is drained by ditching.
**2.** Waterlogged lands. The water table is close to the ground surface and is often flooded during the rainy season. Soils are always water saturated except during dry seasons. Soils have a high organic matter content. Waterlogged lands are common in inland depressions and swamps. Vegetation is water-tolerant shrubs such as those of the heath family, and sedges.
**3.** Submerged lands. The ground surface is under water for most of the year. The depth of water fluctuates from several centimetres in dry seasons to 1 m during floods. The vegetation is very simple, and consists of mostly reeds.

To convert waterlogged lands and submerged lands to economically productive areas, they frequently are terraced. Soil is removed from one part of the area, and piled in another area to make raised areas or terraces. The size of the terraces and ponds varies greatly in accordance with the amount of water, and the agroforestry practices to be instituted. In submerged lands, terraces are smaller and higher and ponds are bigger and deeper than in waterlogged lands. If trees with deep root systems are to be planted, terraces must be sufficiently high. The size of the pond is dictated by the type of fishery to be instituted.

## Selection of tree species

Most wetlands in this region of China are already terraced, but adaptation to moist conditions is still a factor in selecting trees for agroforestry. Tree species should be moisture-tolerant, have a relatively shallow root system, and be able to survive in waterlogged soil for a considerable length of time.

The natural vegetation of this region is a transition from evergreen broad-leaved forest to deciduous broadleaved trees. The following tree species are commonly used for wetland agroforestry practices in this region: water willow (*Salix babylonica*), basket willow (*S. xuchowensis*), long peduncled alder (*Alnus cremastogyne*), trabeculate alder (*A. trabeculosa*), hairy paulownia (*Paulownia tomentosa*), water fir (*Metasequoia glyptostroboides*), swamp cypress (*Taxodium distichum*) and pond cypress (*T. ascendens*).

Of these species, pond cypress often performs best in agroforestry systems. It is a deciduous conifer with a small and thin crown. Its root system is shallow and water-tolerant. It is able to survive with its roots under water for several months up to a year. The species is fast-growing, and its stem is straight and has good wood quality. It sheds its needles in late October, and renews its foliage in April. Thus underplanted winter crops have the benefit of full sunlight.

Other preferable gymnosperms are swamp cypress and water fir. The long-peduncled alder is one of the most preferred deciduous hardwoods. These four species are the most popular for wetland taungya agroforestry in this region of China.

## Selection of crops

In these wetland systems, wheat, barley, oat, oilseed rape and broad bean are commonly planted in late autumn and are harvested at the end of the next spring. Soybeans, cotton, taro and watermelon are seeded in late spring or early summer and harvested in the late summer or autumn. The influence of some of these crops on the growth of pond cypress is given in Table 14.6.

Vegetables such as radish, carrot, and cabbage can be cultivated between the major crops. In some areas, strawberry and mint are also planted. As the trees grow up and their crowns become bigger, light reaching the ground decreases, particularly during the growing season. Shade-tolerant plants or medicinal plants such as ginger, Chinese gold thread, and edible mushrooms are selected for underplanting. In shallow ponds and

**Table 14.6.** The influence of various kinds of interplanted crops on the growth of pond cypress.

| Interplanted crop | Planting date | Initial density (stems $ha^{-1}$) | Survival (%) | Sd Ht (m) | Sg Ht (m) |
|---|---|---|---|---|---|
| Rape, taro, ginger | March 1982 | 1650 | 92 | 1.3 | 2.97 |
| Soybean, oilseed rape | March 1982 | 1650 | 90 | 1.3 | 2.56 |
| Rice, oilseed rape, broad bean | March 1982 | 990 | 85 | 1.3 | 2.56 |
| Cotton, wheat | March 1983 | 1650 | 80 | 1.3 | 1.85 |
| Sugarcane, oilseed rape | March 1983 | 1650 | 55 | 1.3 | 1.65 |
| Lotus root | March 1983 | 1650 | 90 | 1.3 | 2.20 |

Sd Ht, seedling height at time of planting; Sg Ht, sapling height 1 year after planting.

streams, aquatic plants such as rice, lotus, water-chestnut, and arrowhead can be planted as crops. For deeper ponds, water lily, water lettuce, and water caltrop are cultivated as floating crops. Ponds with water depth of more than 1 m are good for breeding fish. There are approximately 40 kinds of fish utilized in fish breeding, including common carp, black and golden carp, and bighead.

## Costs and benefits of the system

If a 2-ha wetland is to be transformed into taungya, 10000 $m^3$ of soil is removed from 1 ha, making it a pond 1 m deep, and is piled on the other hectare, making it a raised terrace 1 m high. The total cost of establishing the system is about US$5500. Of this, US$5000 is for soil removal, US$200 for soil conditioning, US$50 for drainage and irrigation, and US$200 for seedlings and plantings.

One-year-old seedlings of pond cypress are planted with a spacing of 2 × 3 m. Their growth through years 3, 4 and 6 are given in Table 14.7. The economic value of the pond cypress is approximately US$150 per $m^3$.

Cash crops are cultivated just before or after tree planting. Several combinations used in wetland areas of Jiangsu Province, along with their cost/benefit, are given in Table 14.8. The oilseed rape–watermelon–Chinese cabbage combination is the most profitable.

Once the fish pond is established and is managed intensively, a stable yearly income can be expected. A 2-year average is given in Table 14.9.

The annual net profit from the underplanted crops is between US$1050 and US$2750, and US$2090 from the fishes. From year 3 until years 5–6, when the canopy is closing, some profit is still possible from underplanted crops, though production is decreased. Profit from the fishery remains stable, while the potential profit from the trees increases until the end of the rotation.

**Table 14.7.** Growth of pond cypress as a function of age of plantation.

| Age (years) | DBH[1] (cm) | Height (m) | Stocking volume ($m^3$ $ha^{-1}$) | Annual increment ($m^3$ $ha^{-1}$ per year) |
|---|---|---|---|---|
| 3 | 3.31 | 3.60 | 10.20 | 3.40 |
| 4 | 7.10 | 4.36 | 20.40 | 5.10 |
| 6 | 9.08 | 6.00 | 45.30 | 7.55 |

[1]DBH: diameter at breast height.

**Table 14.8.** Costs and benefits from various combinations of underplanted crops in taungya systems[1]. Units are US$ $ha^{-1}$.

| Types[2] | Gross benefit | Expenditure | | | | | Net benefit |
|---|---|---|---|---|---|---|---|
| | | Seed | Manure | Labour | Others | Total | |
| Oilseed rape–soybean | 1375 | 40 | 125 | 150 | 10 | 325 | 1050 |
| Oilseed rape–taro | 2125 | 175 | 200 | 275 | – | 650 | 1475 |
| Oilseed rape–taro–Chinese cabbage | 3175 | 200 | 450 | 500 | 50 | 1200 | 1975 |
| Oilseed rape–watermelon–Chinese cabbage | 3625 | 50 | 400 | 400 | 25 | 875 | 2750 |

[1]First 2-year average after tree plantations established in 1985.
[2]Oilseed rape is seeded in November and harvested in May. Soybean and taro are planted in May and harvested in October. Watermelon is seeded in oilseed rape fields in spring and harvested in July–August. Chinese cabbage is cultivated between two major crops.

**Table 14.9.** Annual costs and benefits from fish ponds – first 2-year average (1985–1986). Units are US$ $ha^{-1}$.

| Species | Gross benefit | Expenditure | | | | | Net benefit |
|---|---|---|---|---|---|---|---|
| | | Baby fishes | Feeds | Electricity | Management | Others | |
| Carps | 4200 | 170 | 1190 | 150 | 450 | 150 | 2090 |

## Pattern III. Chinese Fir + Crops, or Chinese Fir + Tungoil Trees + Crops in Mountainous Areas of the Subtropics of China

The subtropical region of China covers an area of 2.2 million $km^2$, 70% of which is mountainous. The area lies south of the Chinling Mountains and the Huai River, and north of the Tropic of Cancer. Half of the total population of China lives in this region. Because of high population, deforestation is a serious problem. However, conditions for agricultural production in this region are excellent, and farm yields are high as a result of intensive, meticulous cultivation. Taungya agroforestry systems are a major component in food and fibre production.

## Vegetation of the region

In contrast to many parts of subtropical Africa, Central America, and Southwest Asia which have a dry climate, this region of China has higher rainfall, and as a result, the largest area of subtropical forests at this latitude.

The natural vegetation is evergreen broadleaved forests. Trees with oval-shaped leaves and serrate bladed margins are evergreen, and frequently occur on hillsides and in valleys. There are about 1800 genera of flowering plants, almost 60% of the total in China.

This region contains species which are no longer found in other parts of the world. The glaciers of the Quaternary period extinguished many plants that were widespread during the Tertiary. However, the subtropical region of China was spared glaciation, and became a 'sanctuary' for many species.

## Selection of tree species

Commercial forestry is very important in the subtropical region. Bamboo plantations are very common, with hairy bamboo (*Phyllostachys pubescens*) being the most prevalent species. However, the most important commercial timber tree species is Chinese fir (*Cunninghamia lanceolata*), which provides timber for building and other uses.

Chinese fir plantations often are established after harvest of evergreen broadleaved forests, or Mason pine mixed forests. Phosphorus is sometimes a limiting element in the growth of Chinese fir plantations. The species grows best at altitudes of 300–600 m on mountainsides. Productivity of Chinese fir as a function of altitude is given in Table 14.10.

## Selection of short-term economic tree species and crops

There are a variety of crops in the subtropics of China which are used in taungya agroforestry.

**1.** Food crops: wheat, buckwheat, millet, dryland rice, taro, cowpea, sweet potato and cassava.
**2.** Oil-bearing crops: soybean, peanut, oilseed rape, sesame, tungoil tree (*Aleurites fordii*) and cubeb litsea tree (*Litsea cubeba*).

**Table 14.10.** The growth of Chinese fir plantations as a function of altitude and site index.

| Altitude (m) | Average height growth (m $year^{-1}$) | Average DBH[1] growth (cm $year^{-1}$) | Average stocking increment ($m^3$ $year^{-1}$) | Site index (m) |
|---|---|---|---|---|
| 100 | 0.55 | 0.72 | 2.25 | 9.00 |
| 100–300 | 0.73 | 1.07 | 3.75 | 12.00 |
| 300–600 | 0.89 | 1.23 | 6.00 | 14.00 |
| 600–800 | 0.80 | 1.10 | 5.00 | 13.30 |
| 800–1000 | 0.70 | 1.00 | 3.70 | 10.00 |
| 1000 | 0.50 | – | – | 9.00 |

[1]DBH: diameter at breast height.

**3.** Economic crops: tobacco, cotton, ginger, Chinese indigo, ramie and pineapple.
**4.** Medicinal herbs: Dandshen (*Codonopsis pilosula*), large-headed atractylodes (*Atractylodes macrocephala*), purple-flowered peucedanum (*Peucedanum decursivum*), Chinese gromwell, safflower and Job's tears.

Selection of crop is based upon site conditions, local experiences, and social requirements. Millet, corn, buckwheat, and Job's tears ought to be sown in moist and fertile soils. Soyabean, peanut, cassava, sweet potato, sesame and Chinese indigo are more drought resistant, and can be planted on the upper or middle parts of the slope.

## Types of taungya

Chinese fir is usually planted with a spacing of 2 × 2.5 m or 3 m, with a density of 1500–1800 stems $ha^{-1}$. Two general types of taungya are used.

### *1. Chinese fir + crops*

One-year-old Chinese fir trees are interplanted with high stalk crops such as corn. The corn overtops the fir. The second year, millet is interplanted between the fir, and the height of the millet and fir is approximately the same. The third year, soyabeans are interplanted, and they form a low cover beneath the crowns of the fir.

The influence of corn versus rice on the growth of Chinese fir is presented in Tables 14.11 and 14.12.

**Table 14.11.** The effect of corn and rice on the height, diameter and crown growth of 1-year-old Chinese fir.

| Alley cropping | Height (cm) | Ground diameter (cm) | Crown width (cm) | |
|---|---|---|---|---|
| | | | East–west | South–north |
| Chinese fir + corn | 30.80 | 3.83 | 42.75 | 39.40 |
| Chinese fir + dryland rice | 23.43 | 2.72 | 31.08 | 32.72 |
| Chinese fir (control) | 28.67 | 3.66 | 33.75 | 36.30 |

**Table 14.12.** The effect of corn and rice on biomass production of 1-year-old Chinese fir.

| Alley cropping | First year (kg $ha^{-1}$) | Second year (kg $ha^{-1}$) | Third year (kg $ha^{-1}$) |
|---|---|---|---|
| Chinese fir + corn | 456.4 | 1800.7 | 6033.2 |
| Chinese fir + dryland rice | 287.9 | 848.9 | 4161.2 |
| Chinese fir (control) | 255.0 | 1098.9 | 5137.1 |

*2. Chinese fir + Tungoil tree or cubeb tree + crops*

This system works better on more fertile soils. The seeds of the tungoil tree are sown at the same time that the Chinese fir seedlings or cuttings are planted. Usually 3 or 4 fir are planted for every tungoil tree.

It has been found important to select varieties of tungoil tree which are early maturing, have a high yield, and have a short life. The tungoil trees need to be removed at about the seventh year, to reduce competition for the fir.

An overview of the taungya system of Chinese fir with tungoil trees and crops in various regions of subtropical China is given in Table 14.13.

## Conclusion

Taungya agroforestry is a new term for age old practices of combining crop production in tree plantations. These practices have been evolving in China for centuries. Nevertheless, these systems are not viewed as being static. They are constantly evolving, as social and economic conditions change, and as new combinations and new species are found to be more appropriate.

**Table 14.13.** Common combinations of Chinese fir, tungoil tree, and crops in different localities of China.

| Locality | First year | Second year | Third year | Fourth year | Fifth year | Sixth year | Seventh year |
|---|---|---|---|---|---|---|---|
| Jianghua, Hunan | Bu, Mi, or Dr | Cf, Tot, Co | Cf, Tot, Sp | Tos | Tos | Tos | Rem, Tot, Ten, Cf |
| Huitong, Hunan | Cf, Co | Cf, Crops | Cf, Crops | Ten. Cf | Cf | Cf | Cf |
| Jinping, Guizhou | Cf, Mi | Cf, Co, Sp | Cf, Co, Beans | Cf, Sp, Beans | Cf | Cf | Cf |
| Jian Ou, Fujian | Bu, Mi | Cf, Tot, Co, Tb | Cf, Tot, Co, Sp | Cf, Tot | Cf, Tot | Cf, Tot | Cf, Tot |
| Nanping, Fujian | Cf, Mi, or Dr | Cf, Crops | Cf, Crops | Cf | Cf | Cf | Cf |
| Jiangxi Province | Bu, Co, Se | Crops, Cf | Cf, Crops | Cf, Crops | Cf | Cf | Cf |
| Southern Anhui | Bu, Ra | Cf, Co | Cf, Co, Tot | Cf, Co, Tot | Cf, Tot | Cf, Tot | Cf, Tot |
| Longquan, Zhejiang | Cf, Co | Cf, Co, Tot | Crop, Cf, Tot | Cf, Tot | Cf, Tos | Cf, Tos | Cf, Tos |
| Xinyi, Guangdong | Cf, Crops | Cf, Crops | Cf, Crops | Cf, Crops | Cf | Cf | Cf |
| Daminshen Guangxi | Cf, Crops | Cf, Crops | Cf, Crops | Cf | Cf | Cf | Cf |

Cf, Chinese fir; Tot, tungoil tree; Tos, harvesting tungoil seeds; Dr, dryland rice; Co, corn; Sp, sweet potato; Tb, tobacco; Se, sesame; Ta, taro; Mi, millet; Ra, radish; Bu, burning; Rem, removing; Ten, tending.

Modern agronomic techniques such as system modelling can help in predicting the success of new combinations.

## Further Reading


Chung, C. (1978) *An Outline of Chinese Geography.* Foreign Language Press, Beijing, 186 pp.

Editorial (1982) What is Agroforestry? *Agroforestry Systems* 1, 7–12.

Huo, Y.Q. (1981) Water conservation forest is the foundation of ecological balance in mountainous region. *Sino Environmental Science* 1(3), 47–53 (Chinese).

Huo, Y.Q. (1986) Multi-resources, multi-industries and multi-benefits: An integrated system. *Tropical Forestry (Science and Technology)* No. 1, 7–13 (Chinese).

Huo, Y.Q. *et al.* (1986) Studies on hydrological effect and economic evaluation of various forest-soil types in Zhuting Area, Hunan, China. *Intecol Bulletin* No. 13, 53–56.

Huo, Y.Q. *et al.* (1986) Studies on the productivity and water conservation capacity of various forest types in the eastern mountainous region, Hunan, China. *Proceedings of the International Symposium on Mountain Vegetation.* Botanical Society of China, Beijing, pp. 159–163.

Huo, Y.Q. *et al.* (1986) Studies on taking quantitative measurement and economic evaluation of hydrologic-ecological benefits of forest. *Ecological Science* No. 2, 82–89 (Chinese).

Hoekstra, D.A. (1987) Economics of agroforestry. *Agroforestry System* 5, 293–300.

Hsiung, W.Y. (1987) *Types of agroforestry systems for wetland in China.* 12 pp (Scripture).

Labelle, R. (1987) Ten years of work in agroforestry information and documentation. *Agroforestry Systems* 5, 339–352.

Nair, P.K.P. (1985) Classification of agroforestry system. *Agroforestry Systems* 3, 97–128.

Raintree, J.B. (1983) Strategies for enhancing the adoptability of agroforestry innovations *Agroforestry Systems* 1, 173–187.

Von Maydell, H.J. (1985) The contribution of agroforestry to world forestry. *Agroforestry Systems* 3, 83–90.

Von Maydell, H.J. (1987) Editorial: International research in agroforestry. *Agroforestry Systems* 5, 193–195.

Wu, Z.L. (Chief Editor) *Silviculture of Chinese Fir.* Sino Forestry Press, Beijing, 431–440 (Chinese).

# Conclusion

Taungya is a method of reforestation. It evolved as a result of the need for manpower by colonial foresters to carry out planting and weeding activities in tree plantations established to ensure a continuing supply of valuable timber in Southeast Asia. Originally, the only incentive offered to peasant workers was the right to plant subsistence crops between the trees during the first few years of the plantation. More recently, taungya also has been used to alleviate the problem of forest destruction by shifting cultivators. Governments and business enterprises employ the peasants and give them social and economic benefits as an incentive to abandon the tradition of shifting cultivation and to work at establishing and maintaining forest plantations. The social benefits offered to peasants participating in taungya enterprises have increased in some regions to include subsidized housing, and health and educational benefits. As native forests on public lands have decreased in recent decades, governments in a number of Southeast Asian countries such as India, Nepal, Indonesia and Thailand have encouraged systems of taungya that include social benefits.

The system has been partially successful. Large areas of formerly degraded land have been reforested with valuable species. Nevertheless the system is not without problems. Nutrient losses occur during the cropping period, but they are small compared to total stocks of nutrients in the ecosystem. These losses by themselves probably do not affect growth of trees. However, nutrient loss due to clearcut harvesting at the end of each rotation of taungya can deplete stocks to levels insufficient to ensure satisfactory production during subsequent rotations.

Social and economic problems also occur within some taungya systems. A frequent source of discontent amongst the peasants is that they are not able to own property. Some villagers view their participation in the system as only temporary until they can somehow acquire their own land. This results in conflict with the system managers who would prefer the villagers to remain permanently. A nomadic tradition in some indigenous groups also is an obstacle to integrating them into taungya village systems.

Other problems result from competition between trees and crops, and the peasants not having strong incentive to ensure success of the tree plantation. They may have more interest in ensuring good crop growth, from which they will derive personal benefit. Sometimes the peasants deliberately damage the plantation trees in order to favour growth of crop species.

These are generally not problems where trees are grown on land owned by private farmers or local communities. However in such situations there are other problems. They are exemplified by the situation in the Philippines. There, tree production is encouraged on small farms owned by the farmers themselves. Local mills provide markets for the trees. As a result, the problems of land ownership and proper care for trees do not exist, at least for those farmers who own their own land. However, the trees that are planted are fast-growing species such as gmelina that can be harvested for pulp in less than 10 years, a time interval that is reasonable for an individual farmer. Long-lived species such as the Philippine dipterocarps that bring high prices for furniture wood are not planted, because the time required to bring the trees to harvestable maturity is longer than the economic horizon of the individual farmer. Further, the number of peasant farmers is increasing and there is insufficient land for all to acquire ownership.

In plantations of long-lived species such as teak and sal, an important problem is the long time interval during which there is no economic income from the land on which the trees are planted. Because of economic pressure, there is a temptation to cut the trees prematurely. An effort to remedy the problem is being tried in areas of Indonesia. Fruit trees are interplanted among the timber species. The fruit trees begin to produce shortly after grain cultivation ceases, and their economic production can help bridge the gap until the time that the timber trees are harvested. A similar trial has also been initiated in Thailand.

The problem of sustainable yield also can be solved by initiating mixed-age plantations of timber trees. The system could work well after the first trees become large enough to harvest, but there still is the problem of the long interval before that time arrives.

The idea of increasing the diversity of species and of structure in taungya may have an ecological as well as an economic justification. Theoretically, a larger number of species and of age-classes can more efficiently exploit the available resources. More diverse systems increase ground cover and help prevent erosion losses. Organic matter from tree species can contribute to more efficient nutrient recycling in crop species.

As taungya systems become more diversified and serve better the social and economic needs of the local populations, they will contribute toward increasing the sustainability of forestry in general, and timber production in particular. The taungya systems in China are an example of how taungya, over a long period of time, can evolve into a stable form of land management.

The Forest Village Systems of taungya depend upon a workforce of peasant labour. Thus the system would not be appropriate in regions where such peasant labour was not available. It seems to be most appropriate in rural and undeveloped regions of tropical countries, where there is a need for reforestation of valuable timber species and where landless peasants create social and political problems.

Taungya and associated Forest Village Systems are not the perfect answer to sustainable development nor even to sustainable forestry in developing countries. Nevertheless, they are often better than the other alternative for reforestation; forest plantations from which people are strictly excluded. In regions such as Southeast Asia where population expansion and economic pressure in rural areas are common, the social and political pressures are becoming more severe. It is not possible to dedicate large tracts of land solely to production of trees, and to ignore the problems of local peoples. Social and economic problems also must be considered, and local populations must be integrated into the forestry systems.

# Index